FAX POWER

HIGH LEVERAGE BUSINESS COMMUNICATIONS

PHILIP C. W. SIH

VNR VAN NOSTRAND REINHOLD
——————— New York

Copyright © 1993 by Van Nostrand Reinhold

Library of Congress Catalog Card Number 93-23995
ISBN 0-442-01207-1

I(T)P   Van Nostrand Reinhold is an International Thomson Publishing company.
ITP logo is a trademark under license.

Printed in the United States of America

Van Nostrand Reinhold   International Thomson Publishing GmbH
115 Fifth Avenue   Konigswinterer Str. 518
New York, NY 10003   5300 Bonn 3
                                         Germany

International Thomson Publishing   International Thomson Publishing Asia
Berkshire House,168-173   38 Kim Tian Rd., #0105
High Holborn, London WC1V 7AA   Kim Tian Plaza
England   Singapore 0316

Thomas Nelson Australia   International Thomson Publishing Japan
102 Dodds Street   Kyowa Building, 3F
South Melbourne 3205   2-2-1 Hirakawacho
Victoria, Australia   Chiyada-Ku, Tokyo 102
                                         Japan

Nelson Canada
1120 Birchmount Road
Scarborough, Ontario
M1K 5G4, Canada

16 15 14 13 12 11 10 9 8 7 6 5 4 3 2 1

Library of Congress Cataloging in Publication Data

Sih, Phil.
    Fax power : high leverage business communications / by Phil Sih.
        p.   cm.
    Includes index.
    ISBN 0-442-01207-1
    1. Facsimile transmission—Case studies. 2. Business—Communication systems—
    Case studies. 3. Facsimile transmission—Technological innovations. I. Title.
    HF5541.F3S56 1993
    658.4'5—dc20
                                                    93-23995
                                                    CIP

DEDICATION

To my mother, for her steadfast support in the process of writing this book. And to my friends, for their help, encouragement and understanding.

TRADEMARKS

PREFACE

Facsimile (fax) technology and its applications are taking another giant leap forward in utility, functionality, and pervasiveness. The technologies of voice, fax, and computers have been on an intercept course since the late 1980s, and the result has been the emergence of new ways of communicating that promise to bring as many benefits as the adoption of the CCITT Group 3 standard that caused the original explosion in fax use. These new communications methods make use of the fax, the telephone, and the computer to provide faster, lower cost, more convenient, more desirable ways of delivering printed information that cannot be achieved by any other means.

What You Will Learn

Fax Power shows you how you can use the new technologies of fax broadcast, fax-on-demand, and desktop publishing (DTP) to expedite the flow of needed information to your prospects, customers, clients, or colleagues how they want it and when they want it. Fax broadcast allows you to reach hundreds, thousands, or even more people automatically, using fax. With fax-on-demand, you can "dispense" documents automatically by fax in response to a request made by a caller. And DTP-to-fax technology lets you create laser-quality, clear, bold, even dazzling fax documents using computer-based desktop publishing tools and special fax software.

The book starts with examples chosen to let you quickly visualize how your business can use these new technologies. It details actual cases that illustrate the use of these new technologies in publishing time-sensitive news; fulfilling customer requests for documents, service, and support; improving of government service; marketing specialized information, and even helping to prevent crime. It discusses the motivations and benefits behind each application so you can see how you might benefit too. These examples, along with discussions of the technologies and how they work, will give you many ideas for applications.

Those not familiar with the new power fax technologies, or who perhaps do not even how basic fax works, will learn the important aspects of both. You will learn the right questions to ask when considering specific solutions or alternatives. The concepts covered in this book are not product-specific, and are described in generic terms. Thus, as the products and technologies continue to change rapidly, you will be able to increase your understanding on your own and stay current. Some of the material does go into detail, but those who are interested only at the conceptual level can skip ahead and still achieve an understanding adequate for management-level and idea-generation purposes.

Those more technically inclined will find many examples of real products, descriptions on how to work with specific equipment or software, and even discussions of limitations in usage. While this information is by no means exhaustive, it provides enough detail that a technically knowledgeable person should be able to visualize a complete solution and evaluate its technical and economic feasibility. Many tips and shortcuts generally learned only by experience are also provided, which should save considerable time for anyone actually charged with carrying out a project using the products described or similar ones.

The concepts, the examples, and the details covered here will allow you to see how you can do things with fax that most people do not even know can be done. Along with these new possibilities will come new benefits. You will be able to scoop your competition by getting to the customer first, reduce costs while increasing service levels, and automate tasks that most businesses still perform manually. You will see faster, lower-cost alternatives to mass mailings, telemarketing, conventional publishing, and other common business practices that may change the way you do business.

Who Should Read This Book

Who should read this book? Anyone who wants to see actual applications, understand why they are used and how they work, and complete a quick education on a technology of potentially great benefit to nearly any business or profession. You will gain useful ideas and solutions if you work in an area related to:

- Marketing

- Sales

- Customer, Member or Public Service

- Technical Support

- Public Relations

- Office Management

- Communications Technology

- Publishing or Documentation

- Operational Management

- Planning

- Office Equipment or Computing Machinery

Technologists wishing to learn more about fax and its uses will find this book a quick and easy briefing on the subject with an emphasis on applications. It should provide more than enough information for the beginner. Anyone in the fax business (selling fax, servicing fax, promoting fax, writing about fax) who does not know about the new fax technologies should definitely consider this book mandatory reading.

This book is applicable to both the small business and large organization. Departments with well-focused missions may find the ideas presented here of particular interest. This is not a "global think" or "rethink the organization" book. Entrepreneurs will find the high leverage-oriented solutions may provide that extra boost needed to surmount the usual time/money crunch involved in starting a business.

How This Book is Organized

Chapter 1 sets the perspective for understanding the new fax technologies by giving a brief history of fax and a summary-level understanding of how fax works. Fax has an interesting history that can be appreciated by even those already heavily involved in its application.

Chapter 2 profiles organizations that are doing creative and exciting things with fax. You will see what they are doing, why they are doing it, who benefits, and how you could do the same. The chapter also features ideas for uses of the new fax technologies, which can be used as templates and a start for your own thinking. Learning by doing is one of the best ways to get to know about the new fax technologies, and here you have the chance to actually try a few services. The examples of on-demand fax in this chapter are available to any caller.

Chapter 3 goes into detail about broadcast fax, and Chapter 4 covers fax-on-demand. Here you will see how these technologies work and what you need to know to create solutions for your business. Features, equipment, selection, compatibility, hardware, software, and how it all fits together are discussed in these chapters. The coverage of technical details is oriented toward an operational level, and provides the knowledge that will give you the confidence to work with vendors, service bureaus, and consultants.

Chapter 5 covers the use of computer-based desktop publishing (DTP) tools to create digital direct fax documents that are near laser printer quality. This important chapter explains the "hows and whys" of making high-impact, quality fax documents that play to the strengths of fax while avoiding its weaknesses. What are the right typefaces to use? What kind of graphics look good? What programs do I need? Questions such as these are addressed. Hard technical facts are provided when they can help you avoid potential stumbling blocks and understand critical processes. This chapter will be of central importance to anyone who must create or direct the design of documents for fax transmission.

Performance issues are covered in Chapter 6, which shows ways to minimize costly transmission time, reduce transmission errors, improve reliability, and avoid problems that can degrade the appearance of a fax. You will learn ways to save money by trimming the cost of practically any fax transmission, whether from a fax machine or computer-drive fax modem.

Chapter 7 examines some technology trends and how they might affect fax in the future. This look at tomorrow takes a near-term view which should be useful for immediate planning purposes, as well as a longer-term focus, more suited to wide-ranging strategic thought.

Finally, the Appendix is resource directory which should save you time in tracking down vendors of both products and services, and also lists some other sources of information.

A Word on Price Information

It is virtually certain that some of the products, services, and prices discussed here will change before this book is in print. This information is included because it provides a basic understanding of relative cost, which is important to evaluating application alternatives. You should have an approximate idea whether something costs a dollar, ten dollars, a hundred, or a thousand. Therefore, price information in this book is should be used only as a basis of relative comparison as of mid-1993. Keep in mind that the price of electronic technology has been steadily declining and can be expected to do so at a similar rate in the near future. Also remember that street prices (what you might actually pay) often are lower than listed prices, sometimes by 30 to 40 percent. You should consult periodically published market sources such as the *Fax Buyer's Guide* for current price data. (See the Appendix for contact information.)

ACKNOWLEDGMENTS

Dianne Littwin, Senior Editor at Van Nostrand Reinhold for her patience and willingness to take a chance.

Rusty Weston, Senior Writer at *Corporate Computing* magazine, for help on the initial proposal and getting me into this project in the first place.

W. David Schwaderer, friend and author of many technical books, for advice and encouragement as an author.

TABLE OF CONTENTS

Chapter 1

A Brief History of Fax

Facsimile and its use has always been at the confluence of many technologies, and in some cases, has served as a strong motivation for advancement in the state of the art or invention in a number of fields. The new facsimile applications discussed in this book are based on a combination of digital image, computer and telephony technologies. In following chapters we will be looking in detail at these new applications, how they work, and how you can benefit from using them. But before we get into that material, this brief historical perspective on fax should help you better see how fax got to where it is today.

The first officially recognized fax was patented in 1843 by a Scottish physicist named Alexander Bain. He called his invention a "recording telegraph" as it was based on the telegraph technology of that time. Sending and receiving machines were connected to either end of telegraph wires. The sending machine sensed "dark" and "light" spots using a stylus attached to a pendulum that passed over metal type. The metal-to-stylus contact was either open or closed depending on what was on the metal plate. This was used to regulate a direct current sent over the telegraph lines. The receiving machine recorded these spots on paper using a similar stylus attached to a pendulum. The pendulums were driven by a clock type mechanism were kept synchronized by the use of electromagnets and "sync" pulses.

From that time until the early part of the 1900's fax continued to use telegraph technology because at first there were no telephone lines, and later because telephone lines could not communicate over the longer distances of telegraph lines. Various point-to-point fax services were established, mostly to provide support for news applications between large cities.

After 1900, as telephone service began to expand into what we now have today, fax moved from telegraph lines to telephone lines. Between 1900 and about 1940, fax continued to see highly specialized application. It was still mainly used to send news photos for specialty news services. It was during this time period that portable, self-contained fax transmission equipment was first developed, moving fax from reliance on expensive dedicated transmission facilities, to being able to connect to practically any telephone line.

In the 1940's the U.S. military began to use fax for several purposes including field operations and weather maps. During that time, the U.S. Air Force set a worldwide standard for transmission of weather maps via fax. In the years following World War II, many specialized fax services developed for news, weather, and law enforcement. Use of fax then started to expand into business offices.

As with most early technologies, fax equipment from different makers was not at all compatible. There were no standards and each maker used a different scheme for their fax designs. When fax was used mostly for specialty purposes, this did not pose a problem. But as businesses began to use fax, the need for compatibility between different equipment started to emerge.

In 1966 the Electronic Industries Association (EIA) a U.S. standards organization issued its first standard on fax, Standard RS-328. Using this standard, diverse fax equipment could for the first time communicate. Compatibility was slow in coming, and in 1968 the CCITT, an international standards group issued its recommendation (they don't call them standards) on Group 1 fax. Group 1 offered fax compatibility outside of North America, the domain of the EIA. But standards alone were insufficient to move the fax market into the mainstream. Several other key enablers still did not yet exist.

One key enabler for the explosion of the fax market was *direct* connection of non-telephone company equipment to the public telephone network. Several events eventually led up to this. The first event was the "Carterphone" decision of 1967. That was a court case where a man named Carter won a court case against AT&T that opened way for *acoustic* or *inductive* coupling (but not yet *direct* coupling) of equipment to the public telephone network. At that time, all telephone services, both local and long distance in the U.S., were provided by AT&T as a government sanctioned monopoly. Before that time, connection of non-AT&T equipment to the telephone network was only permitted through expensive, AT&T supplied specialized fax interface equipment. The ostensible reason for this was to protect the telephone network from potentially harmful non-compliant signals.

In 1969 fax machines (as well as other types of equipment such as data transmission equipment) could be *electrically* connected to the telephone network, but again, not *directly*. This was done through the notorious *data access arrangement*, another expensive, cumbersome piece of specialized equipment. But this was significant, since it now eliminated noise from acoustic or inductive coupling and provided a way for the fax to perform the tasks of automatically dialing to originate a call and detecting ringing answering.

Finally, in 1978, the Federal Communications Commission, the rule-making authority for telephone services in the U.S., issued Rules and Regulations Part 68, paving the way for *direct* connection of equipment (including fax) to the telephone network. By this time, the CCITT had already issued its Group 2 recommendation (1976) which was an improvement both in performance and cost over Group 1.

Another key enabler was the availability of low-cost, mass manufacturing. Around the same time as the Group 2 recommendation, the Japanese manufacturers began to get into the market for facsimile equipment by aggressively manufacturing and selling Group 2 equipment at reduced prices. Their early interest and constant work in the area of standard facsimile continued through the fax explosion and is probably the main reason they are the dominant makers of fax equipment worldwide today.

Two more enablers, both related, still needed to appear on the scene before fax would really become popular. The first of these was the development of digital facsimile, a technology dependent on the development of what was then called "high-speed" datacommunications modem technology and the second was development of integrated circuit technology. Without the modem technology it would not be possible to send data at rates sufficient to compete favorably against the existing Group 2 equipment, and without integrated circuit technology, it would not be practical to build digital equipment sufficient to handle the demands of fax.

In the mid 1970's the CCITT began work on its recommendation for Group 3 fax, but it was not until 1980 that it finally issued it. During this time there was extremely rapid development in both integrated circuit and modem technology, and of course the continued interest on the part of the large Japanese manufacturers. By the time the recommendation was issued, the technology, manufacturing, and marketing were all in place to touch off the explosion in the fax market we are still seeing today.

Between 1980 and now, mid 1993, most industry sources estimate the total number of Group 3 fax machines in the U.S. alone has grown to over 20 million. Worldwide communication between businesses, and increasingly between individuals, is now not only possible but expected. It is estimated that over 10 billion fax messages are sent each year. Fax has virtually put telex service out of business, having pushed it into its last remaining market, less developed countries where phone service is unacceptably poor. (Remember the telegraph lines and fax?) Fax accounts for more than half the telephone line traffic between the U.S. and Japan.

Another wave of the fax explosion has started based on the integration of computer, telephone (voice) and fax technologies. Part of this is driven by the constantly declining cost and increasing sophistication of computing equipment, especially the personal computer. Another trend is the increasing acceptance on the part of the public that automation, when properly done, can provide a great convenience and benefit. This has led to an increase in the use of automated voice technologies such as automatic attendants, voice mail, and interactive voice response. Finally, the Group 3 standard has itself been evolving and improving with the available technology. This is possible because of the extensibility designed into the recommendation, a feature that recognizes that future developments will need to be incorporated and remain compatible with existing products. The result has been that the higher transmission speed of 14.4kbps was added to the Group 3 recommendation in 1991 along with error correction (ECM) and data higher compression (MMR) techniques. These are covered later in this book.

These new computer-based fax technologies along with the now widespread use of the fax machine open up stunning new possibilities. If you think of each fax machine as a remote printer to which you can send computer originated documents or as a remote terminal which can be used to interact with a computer from practically anywhere in the world, then you can begin to see what's possible. In the chapters that follow, we are going to examine in detail how businesses and other organizations today are using the new fax technologies and how you can too.

CHAPTER 2

FAX POWER IN ACTION — REAL EXAMPLES OF FAX BROADCAST AND FAX-ON-DEMAND

We're going to examine several examples of actual broadcast fax and fax-on-demand applications whereby organizations, both large and small, are using the new fax power technologies to meet their information and communications needs today. The best way to get started is with a real example. If you have a fax machine, you can see for yourself how fax-on-demand works right now. Using the handset on your fax machine, dial 415-771-7400, listen to the message, and then press the START button on your fax machine. It's that simple.

Now that you have seen how fax-on-demand works, let's look at a few more examples of both fax-on-demand (also called fax response or fax-back) and broadcast fax. We will review several examples in-depth. For each we will ask:

- What are these people doing and how are they doing it?

- What information is in the fax?

- Who is using this information and why?

- What equipment, services, and techniques are used?

- What are the other benefits or factors involved in the use of fax technology?

Here are the cases we will examine:

7

McGraw-Hill, one of the largest publishing firms in the U.S., has more than thirty *fax-based publications.*

Io Publishing delivers a unique *daily fax newspaper* to a select and highly influential group of investors and executives in the biotechnology industry.

Intel Corporation, the well-known maker of computer components, add-on products, and software, uses on-demand fax to provide both *sales support* and a unique form of *customer support.*

Borland International, a large software producer, provides detailed *technical reference* information to customers through a virtual *library of fax documents*.

The Foresight Institute, a small membership organization, uses fax-on-demand to provide a *clearinghouse for documents* of interest to its members.

The U.S. Navy/National Oceanographic and Atmospheric Administration Joint Ice Center uses fax-on-demand to provide instant access to the *latest charts showing ice flows around the world.*

The Ottawa Police CrimeFAX helps stop credit card fraud using a *unique desktop publishing and fax broadcasting setup.*

Now let's go to our first case.

McGraw-Hill

McGraw-Hill believes in fax and has for a long time. The company is on the leading edge of fax technology applications and is always looking for new markets to serve with the information and editorial content it produces. For McGraw-Hill, fax technology has played a key role in more than thirty publications to date, some of which involve unprecedented business partnerships. For example, *Standard and Poor's,* a division of McGraw-Hill, has launched a fax service with the Los Angeles Times Syndicate and the Prudential Insurance Company. Subscribers to the Los Angeles Times and the San Francisco Chronicle can receive up-to-the-minute information by fax on as many as fifteen stocks at the end of each business day.

You can get a quick perspective on this extensive fax publishing operation by looking at Table 2-1.

Let's take a look at one of these publications, *Tanker Fax* and see how it works. *Tanker Fax* (see Figure 2-1) is written and produced each work day in London, England, at a branch office of McGraw-Hill. There the staff gathers the information that will go into each *Tanker Fax,* and then creates the layout for the two-page fax document using Ventura Publisher software by Xerox Corporation. After the layout is complete and ready to go to press, Ventura Publisher is used to generate a PostScript format file containing the document. That PostScript file is transmitted electronically to COMwave, a fax transmission service bureau in London. COMwave converts the PostScript file to a digital direct fax document and then transmits *Tanker Fax* through its international fax broadcasting network to subscribers around the world. Figure 2-2 illustrates this process.

The value of *Tanker Fax* and its timely information is so great that subscribers are willing to pay $2,000 to $3,000 a year for it, depending on their location on the globe.

In deciding whether to do the fax transmission itself or employ a service bureau such as COMwave, McGraw-Hill considered the largest factor to be its desire to focus on its business of gathering and providing information. The firm does not see any strategic benefit in taking on the operations of transmission, and has been quite satisfied with the cost and services of COMwave.

Fax is a new distribution medium for *Tanker Fax,* which started out in the early 1980s using Telex for transmission, was slow compared to fax. Telex is only capable of transmitting information at 110 bits per second (bps); fax transmits at 9,600 bps. While this is not exactly a completely accurate comparison, it does

PUBLICATION	DELIVERY	DESCRIPTION
Addenda Fax	on-demand	U.S. Building project plans and specifications. Additional to similar information available on microfilm.
Product Line	on-demand	New product specifications for architects and design professionals. Augments Sweet's regular newsletter published ten times per year.
KennyAlert	broadcast and on-demand	Expands market for JJ Kenny's securities information to smaller brokers and individuals.
Blue List	broadcast or on-demand	Bond prices, yields, maturity dates and other information on approximately 9,000 daily corporate and municipal bond issues. Used by mid-sized brokers and individuals.
Wall Street by Fax	broadcast	Two-page S&P stock report covering any of the approximately 4,100 stocks listed on the NYSE, ASE, and NASDAQ.
Fact Set	broadcast	Recommendations on about 140 initial public offerings a year.
Research Reports	on-demand	Database-derived, seven page reports on public firms. Used by brokers and investors. Include analysis and historical data updated daily.
Financial Fax	broadcast	Daily reports on up to fifteen stocks. Published at close of each business day; shows highs, lows, volume, and other performance information.
Credit Wire by Fax	broadcast	A derivative of the Credit Wire on-line computer database. Provides ratings alerts to investors on approximately 5,000 credit issuers.
SprintMail	broadcast	Daily ratings and changes on corporate and municipal bonds to portfolio managers, brokers, banks, and news agencies.
Forecast of U.S. Economy	broadcast	Analysis of events affecting U.S. economy. Several hundreds subscribers in U.S.
Special Fax	broadcast	Reports on economic impact of events from budget votes and natural disasters.
Energy Price Outlook	broadcast	Monthly projections of oil and gas prices, market developments.

Table 2-1: McGraw-Hill Fax Publications Summary

PUBLICATION	DELIVERY	DESCRIPTION
Foreign Exchange Service	broadcast	Monthly currency advisory service for corporate treasurers and planners. Examines economic and political changes in thirty two countries.
DRI Financial Fax	broadcast	Published each Friday. Examines financial activities of past week. Includes charts and graphs plus calendar of government economic releases and treasury auctions.
Economic Fax Service	on-demand	Value-added market and economic information to investors and global money, bond, currency, and equity traders.
Gaswire	broadcast	Daily overnight fax on bid offerings for natural gas contracts.
Metals Price Alert 1	broadcast	Daily transmission at 1500 hours London time. Latest Asian-Pacific and European metals market information, London Metals Exchange results, opinions.
Metals Price Alert 2	broadcast	Daily transmission at 1630 hours New York time. European and North American market information and Comex, NY Mercantile, and NY dealer quotes.
Metals Week Price Notification Service	broadcast	Daily, weekly, or monthly updates of 450 Metals Week proprietary prices and exchange-traded prices.
Marketscans	broadcast	Daily summaries of prices, market activity. Three editions for Europe, Asia, and Arabian Gulf.
Oilgram News/Wire	broadcast	Daily compilation of petroleum-related news.
Crude Oil Marketplace	broadcast	Crude oil market news.
Feedstock Report	broadcast	Weekly report on oil by-products market.
Bunkerwire	broadcast	Twice weekly report on marine fuel prices and supply.
Tanker Fax	broadcast	Daily updates on worldwide tanker market.
Petrochemicalscan	broadcast	Weekly update on petrochemical spot prices.
LP Gas Wire	broadcast	Liquid petroleum spot market prices and information.
OLEFINscan	broadcast	Spot and contract prices for a specific oil by-product.
Polymerscan	broadcast	Weekly price and event news on bulk polymer market.
Solventwire	broadcast	Periodic reports on the solvent market.
Afternoon Executive News Summary	broadcast	Daily summary of general news for corporate executives. Includes advertisements. Done with television station WRTV of Indianapolis.

Table 2-1 (continued): McGraw-Hill Fax Publications Summary

present an approximate difference and illustrates one of the major reasons fax has eclipsed Telex. Telex also places many limits on the design of a publication because it is able to print only in uppercase characters at a maximum of eighty per line. It is interesting to note that Swift Global Communications (see the Chapter 3 section on service bureaus) which started out in the Telex business in 1978 has moved almost entirely to fax broadcasting.

In 1988, McGraw-Hill began transmitting *Tanker Fax* by fax instead of Telex. And as you can see from Figure 2-1, it is a vast improvement in design over what could be done with Telex. But design was not the only motivation that drove McGraw-Hill to begin using fax; cost and market demand for convenience were also factors.

All of the information contained in the publications listed in Table 2-1 is also available through an on-line computer database, and much of it is still available via Telex as well. In comparison with using an on-line service, receiving a fax is much more convenient and involves no time on the part of the subscriber. Users of the on-line databases must first obtain and learn to use the proper computer and datacommunications equipment. Once past that barrier, they must take the time each day to access the system and obtain the information they want. Fax does away with all these procedures by simply delivering a daily digest of the information a subscriber is most likely to want. No special equipment or training is necessary, as most subscribers already have fax machines.

For nearly all international communications, fax has replaced Telex. Telex is far more expensive than fax, and has a complicated rate structure. It requires that special terminal equipment and leased lines be installed at each subscriber location. By using fax instead, McGraw-Hill saves greatly on cost while improving the overall appearance and quality of the product.

Not all of the McGraw-Hill publications shown in the table are produced using desktop publishing. In fact, the majority are just digital direct faxes of ASCII text, but McGraw-Hill expects that more will be moving toward desktop publishing now that the technology and methods have

Platt's Clean Tanker *fax*

Tanker intelligence from the leading oil market specialist   Vol.11 No.61 Tuesday 30th March 1993

Market Latest (130 pm GMT): May Brent talked at $18.75/bbl. Apr Gasoil talked at $174.75/mt.

Market Fixtures

Vessel				Route	Rate	Charterer
BURWAIN PACIFIC (S)	58	CL	APR3	RAS TANURA - BRAZIL	W126	PETROBRAS
ICARUS	28	CL	MAR30	CROSS AG/AG - INDIA	$127K/W150	BANOCO
CAPT.X.KYRIAKOU	25	CL	APR	YANBU - EAST	W190	CNR
PRIMA MAERSK	37	GO	APR6	SINGAPORE - HONG KONG	$185K	SIETCO
BP ADMIRAL	35	CL	APR8	GUAM - S.KOREA	$225K	KANEMATSU
WORLD BRIDGE	35	UL	MAR	SINGAPORE-JAPAN	W143	ITOCHU
PRIDE	30	CL	APR3	S.KOREA - JAPAN	$150K	CNR
DIYYINAH	30	CL	APR9	S.KOREA - JAPAN	$130K	TAIYO
UNITED SELMA	60	CL	APR20	CONT - KHARG	RNR	CARGILL
GORGONA	37	CL	PPT	CONT - EC/WC INDIA	$800/725K	CNR
PETROBULK TBN	30	UL	MAR	ANTWERP - US	RNR	PETROFINA
LOUISE	28.5	UL	APR	BILBAO - TA	W185	BP
MOLLER 'R' TYPE	22	LD	APR3	MONGSTAD - MED	W180	STATOIL
TORM KRISTINA	60	CL	APR20	WEST MED - JAPAN	$1.35M	FLEET
NORDTRAMP	55	CL	APR8	ARZEW - USG	W117	LYONDELL
KRITI COLOR	35	CL	APR	BLACK SEA - MED	W140	CNR
ANTONIO D'ALESIO	30	CL	APR	SKIKDA - UKCM	W160	FLEET
ROSSI	26	NA	APR3	SKIKDA - UKCM	W175	BASF
BURWAIN CASTOR	25	NA	APR1	LIBYA - UKCM	W180	ELF
NERVI	23	GO	APR2	RAS LANUF - TOBRUK	$70K	NOC
URZHUM	29	CL	PPT	T/C 1/1 MONTH	$12K/D	MARAVEN

Daily rate assessments

FROM: TO:	UKC	MED	BSEA	INDIA	SING	CHNA	CARIB	USWC	RSEA	AG 30,000	AG 50,000
UKC	160	160	160	450K	NRR	—	200	—	525K	495K	*120
MED	150	150	150	350K	NRR	—	200	—	405K	395K	*122
AG	700K	550K	550K	350K	255K	—	900K	RRR	270K	120K	*180K
INDIA	650K	600K	600K	—	140K	—	900K	1.4M	150-	150-*NRR	
SING	750K	790K	790K	400K	47K	165K	900K	650K	150-	150-*120	
JAPAN	1.2M	1.0M	1.0M	630K	170	—	895K	550K	140-	140-*120	
USWC	—	—	—	630K	NRR	350K	NRR	—	825K	1.0M	*NRR
USAC	175	170	170	550K	NRR	—	215	RRR	525K	500K	*120
USG	175	170	170	550K	NRR	—	215	RRR	575K	540K	*120

Editorial Note: Platt's will be making the following changes to daily clean tanker rate assessments on 1 April 1993. 1) Replacing current 50,000mt rates from the AG with 55,000mt. 2) Replacing the current 50,000mt UKCont/Med/USAC/USG assessment from Worldscale to Lumpsum.

Regional price differences

Regional price differentials computed from Platt's prices and converted into Worldscale.
( ):reverse voyages, +/-:up/down today, X:change of profit direction.

		$/MT	WS			$/MT	WS
NWE-USAC	Mogas	(12.7)+	(203)+	NWE-USG	Mogas	4.6-	55-
	Gasoil	9.4-	150-		Naphtha	20.7+	247+
	LSFO	5.7	91		Gasoil	(4.2)-	(50)-
	HSFO	6.5	103		LSFO	(2.4)	(29)
MED-NWE	Mogas	0.0	0	MED-USAC	Mogas	(16.7)+	(222)+
	NAPHTHA	14.0	237		Gasoil	12.9-	171-
MED-NWE	Gasoil	10.8	217	MED-USAC	Jet	13.5-	180-
	Jet	15.5	313		HSFO	8.0	106
	HSFO	6.0	121	NWE-MED	LSFO	7.5	137
CAR-NWE	Naphtha	14.5+	218+	CAR-USAC	Gasoil	18.7+	561+
	Gasoil	16.9+	254+		Jet	17.9-	536-
	Jet	21.9+	329+		HSFO	9.0	268
	HSFO	8.5	128	CAR-USG	NAPHTHA	42.2+	1142+
CAR-JAPAN	Jet	47.2+	320+	AG-MED	NAPHTHA	(4.9)+	(55)+
MED-JAPAN	Jet	35.8	227	AG-JAPAN	NAPHTHA	15.0	136
SPR-JAPAN	Jet	11.5	193	AG-SPORE	Jet	13.1	210

IMO chief says world shipping standards under threat

An aging, poorly maintained tanker fleet, staffed by inadequately trained and inexperienced personnel has undermined the world shipping industry's ability to maintain the standards necessary for safe operations, according to William O'Neil, the secretary-general of the International Maritime Organization. Speaking at the 1993 International Oil Spill Conference, O'Neil said the industry has been in the doldrums since the boom years of the 1970s and shipowners, therefore, have been disinclined to order new ships. 'If there has been a chance of making a profit it has been by squeezing out more life from the existing fleet without risking the large capital outlay required for new buildings,' he said. 'The result is that the world fleet is getting older, with the average age of a tanker now...15 years.'

Brisk Med-Far East Trade

Brisk Med-East trade continued at steady rates as *Torm Kristina* fixed to *Fleet* on 60,000mt from the west Med to Japan at $1.35-mil. Movement to both east and west out of the UKCont was also noted at firmer rates as **BP** fixed *Louise* to the US at w185 on 28,500mt. Traders in the AG LR market remained inactive in the face of slow end-user demand in Japan. **Petrobras** reamins active, however, and failed *Omegavenure L* to Brazil. The vessel was replaced with *Burwain Pacific* which fixed on subjects on 58,000mt at w126 to Brazil. Brokers in the Far East reported relatively active trading today although some rates softened as a result of the continuing plague of surplus tonnage. **Taiyo Oil** were heard to have fixed *Diyyinah* on 30,000mt from South Korea to Japan at a soft $130,000.

Oil News

- *Nepal has issued a new tender to buy 30,000 tonnes of superior kerosene (SKO) and 30,000 tonnes of 1% sulfur content gasoil, both for delivery April 20-30, oil trading sources say. The tender closes April 5 at 2400 Nepal time (1830 GMT) and is valid until April 7. Nepal will take the kerosene cargo at the port of Goa/Kandla on the west coast of India and lift the gasoil cargo on a C+F Madras/Hladia basis, in the east of India.*

- *Venezuelan union leaders negotiating a new contract for 65,000 oil industry workers last Friday began a legal countdown that by this Wednesday could lead to a series of work stoppages, and potentially, a nationwide oil industry strike. The action by leaders of Fedepetrol and Fetrahidrocarburos, giving the oil industry 120 hours from last Friday to settle differences with the unions before beginning localized strike action, is the first of its kind since PDVSA was nationalized 17 years ago. An oil strike could shut down a majority of PDVSA's operations, according to industry observers.*

Editors: London; Daniella Gluck, Tel 44 81 545 6104, New York; Allyson LaBorde, Tel 1 212 512 4532. Tokyo; Tel 813 3593 9803.
Sales: London; Carmel Kerrigan,Tel 44 81 545 6145. Singapore; Josephine Soh. Tel 65 532 2800. New York; Dawn Soares. Tel 1 212 512 6019

Figure 2-1: Digital Direct Fax Newsletter: *Tanker Fax*

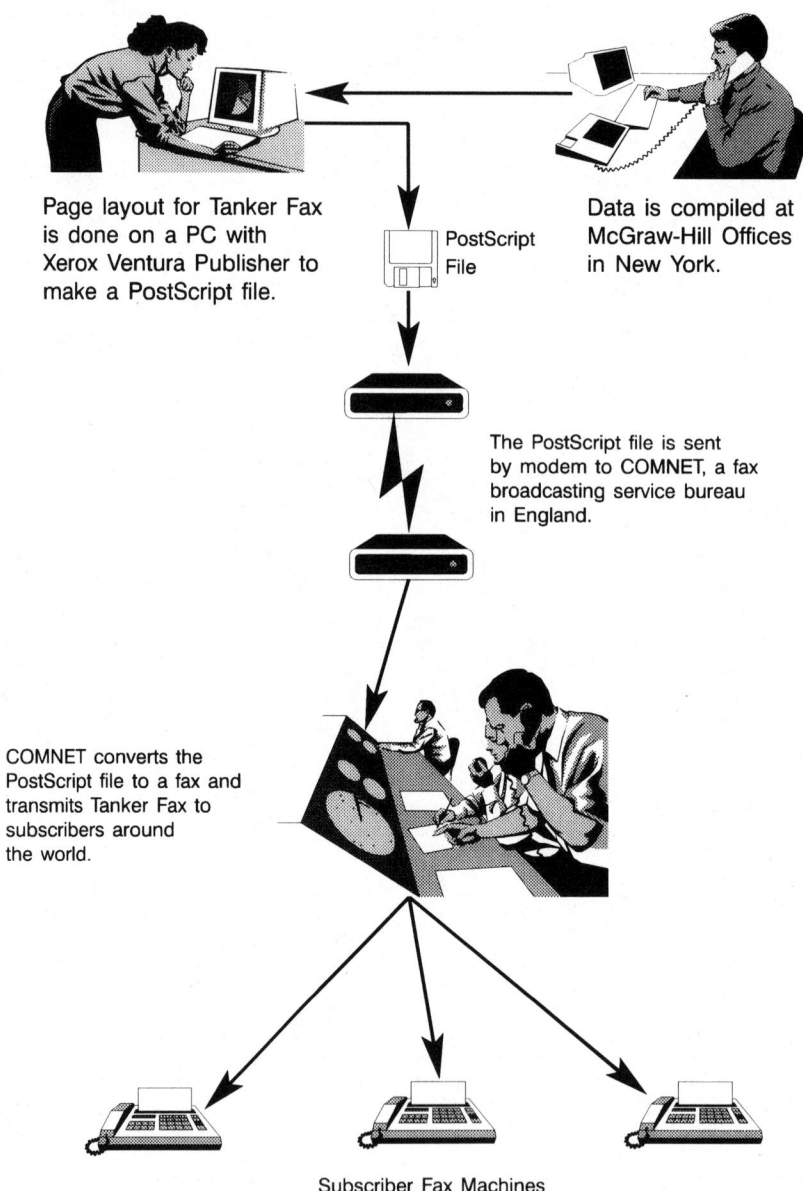

Page layout for Tanker Fax is done on a PC with Xerox Ventura Publisher to make a PostScript file.

PostScript File

Data is compiled at McGraw-Hill Offices in New York.

The PostScript file is sent by modem to COMNET, a fax broadcasting service bureau in England.

COMNET converts the PostScript file to a fax and transmits Tanker Fax to subscribers around the world.

Subscriber Fax Machines

Figure 2-2: Tanker Fax Production and Delivery

become familiar. Another coming trend is the augmentation or conversion of traditional print publications into fax format to improve timeliness and lower costs.

Intel FaxBACK

Intel Corporation is a large multinational manufacturer of integrated circuits, printed circuit board products, and specialized computer systems. The company is best known for its microprocessors (like the i386 and i486) which are the central component in most personal computers today. But Intel produces much more than microprocessors; the number of different products it sells is in the high hundreds if not in the thousands. Nearly all the products are highly technical in nature and require plenty of documentation as part of the sales and customer support process.

Since 1989, the Personal Computer Enhancement Division (PCEO) of Intel has used a sophisticated yet crisp and easy to use on-demand fax system to help meet the needs of its customers for specifications, product literature, price lists, technical bulletins, and even software program files and other computer files.

This was one of the first fax-on-demand systems developed, and it has an interesting history. *Intel FaxBACK,* (see Figure 2-3) was started as an after-hours project by Art King, Dan Wagner, and a few other enterprising members of the Intel staff just shortly after the company introduced an early version of the Intel SatisFAXtion fax and data modem. King comments that "Intel came out with this product and we all knew what it could do." They set about building their first fax-on-demand system, and, not surprisingly, used the Intel SatisFAXtion fax modem in it.

Using *Intel FaxBACK* is straightforward, and the consideration that was put into the structuring of the information and scripting of the voice prompts is evident. Callers use a touch-tone phone to dial 800-525-3019 and are greeted with instructions and a voice announcement of important product news. The instructions tell callers that they can bypass the message at any time by pressing any key to get to the main menu. The main menu is clearly identified with a brief recorded message. Most systems do not tell you where you are; but this is important and should be done in every case.

The number of main menu selections is limited to five, and is structured so the most likely choices come first. Of great convenience is the point in the voice menus that allows customers to order catalogs that list the available documents. This choice is available from the main menu. It is good design practice when using fax response to provide lists and directories by fax rather than extensive voice menus. It is much easier to find what you want in a printed list than it is to try to listen to and recall a complex set of choices.

Each of the handful of catalogs addresses a specific product area. Once a catalog is ordered, selecting a document for delivery by fax is easy. Document ordering is the first item in the main menu and documents can be entered by number in rapid succession.

Entering your fax number completes the order, and within seconds a return call is placed by the *Intel FaxBACK* system to begin the transmission of your documents.

One of the first things you will notice about the return fax as well as the voice prompts is their commercial orientation. Ever enterprising, Intel wastes no time or space, and places its own brief, tastefully done advertisements on the cover sheets and in some of the voice menus.

Intel FaxBACK is one of few systems that uses pleasant beep tones to give callers a strong sense of timing in the interaction and some quick and intuitive feedback. Each recorded message is followed by a brief set of tones that seem to be a natural way of telling callers what to do and when to do it. Table 2-2 lists these tones and their apparent meanings:

To: 14157760615   From: FaxBACK(tm) Intel Support #71   1-17-93 2:02pm   p. 1 of 6

FaxBACK™ *Information from Intel's Automated Customer Support Service*

New–OverDrive™ Processors for i486™ DX systems!

Upgrade your Intel486 DX system with
an Intel OverDrive Processor and run
all your software up to 70% faster!

For details, please order document
number 3006.

To: 14157760615   Date: 1-17-93

From: FaxBACK(tm) Intel Support #71   Page 1 of 6

Please deliver this Fax to the person at this phone number who requested it.

Thank you for using FaxBACK!

FaxBACK can give you the latest information about all of our PC and LAN
Enhancement products 24 hours a day. You can receive Intel product
brochures and price lists, installation instructions, troubleshooting
information, extensive compatibility notes, and much more.

Here's how you can reach us at Intel:

Product Information: (Monday-Friday, 7AM-5PM Pacific Time)

All product lines   800-538-3373   or   503-629-7354

Automated Support: (24-hours a day, every day)

```
FaxBACK                 503-629-7576   or   800-525-3019
Fax                     503-629-7580   or   800-458-6231
Bulletin Board Service  503-645-6275   (8-N-1, up to 14.4 KBPS)
CompuServe              GO INTELFORUM  (7-E-1)
```

Figure 2-3: *Intel FaxBACK* Cover Page and Typical Response

To: 14157760615   From: FaxBACK(tm) Intel Support #71   1-17-93 2:03pm   p. 2 of 6

```
┌─────────────────────────────────────────────────────┐
│        FaxBACK(TM) PC & LAN Enhancement Products      │
│             Product Documentation Catalog             │
└─────────────────────────────────────────────────────┘
```

The following pages have an up-to-date listing of the documents available from FaxBACK for the product line you asked for. With this catalog you can order product brochures and price lists, extensive compatibility notes (the same information Intel's Customer Support uses), installation and troubleshooting instructions, and more. Simply call FaxBACK and enter an "Express Order" for the documents you desire. Before you call, have the phone number of your Fax machine or Intel Fax board or Faxmodem handy.

You can reach FaxBACK at (503) 629-7576 or (800) 525-3019, 24-hours a day, seven days a week. In Europe, please call +44-793-432-509.

For your convenience, this page has document numbers for Data sheets and general information about all our PC & LAN products and services.

Thank you for using FaxBACK!
-- Intel PC & LAN Enhancements
-- Customer Satisfaction Staff

General Information
═══════════════════════════════════════
FaxBACK Catalog: What's New Since
   the Last Printed Catalog . . . 9010
Intel PC and LAN Enhancement
   Price List 9000
Mobile Computing Standards . . . 3420

CPU Upgrades (Boards)
═══════════════════════════════════════
CPU Enhancement Product Overview 9002
SnapIn 386 9208

Fax & Modems
═══════════════════════════════════════
Fax and Modem Product Overview . 9004
SatisFAXtion Products Guide . . 9406
14.4EX Modem 9434
Hand Scanner for SatisFAXtion . 9412
Intel 2400 bps V.42bis Modems . 9430
Intel 9600EX Modem 9432
Intel FAXability Software Guide 9418
SatisFAXtion Modem/100 9414
SatisFAXtion Modem/200 9415
SatisFAXtion Modem/400 9416
SatisFAXtion Modem/400e 9417

Math & CPU Processors
═══════════════════════════════════════
Performance Upgrade Components . 9003
Intel Math CoProcessors 9300
Intel OverDrive Processor . . . 9306
Intel OverDrive Feature/Benefits . 3018
RapidCAD Engineering CoProcessor . 9302
Intel's 32-Bit Microprocessor
Architecture Overview 9013
Intel iCOMP Performance Index . . 3202

Memory Enhancements
═══════════════════════════════════════
Memory Enhancement Products . . . 9001
Above Board ISA 9103
Above Board MC 9102
Above Board ISA I/O Module 9111
Above Board Plus & Plus 8 9100

Network Adapters & Hubs
═══════════════════════════════════════
Network Adapter Product Overview . 9006
EtherExpress 16 Data Sheet 9620
EtherExpress 32 Data Sheet 9621
EtherExpress MCA Adapters 9624
EtherExpress TPE Hub Data Sheet . 9623
EtherExpress TPE Hub: Novell's
   NetWare Hub Technology Overview 6502

Network Management & Utilities
═══════════════════════════════════════
Network Management Products . . . 9007
Intel Guide to Network Utilities . 9579
LANDesk Manager 9555
LANProtect 9558
LANSight Support 9550
NetSight Analyst 9551
NetSight Professional 9553
NetSight Sentry 9552
Node Management Strategies 9009

Network Print & Fax Servers
═══════════════════════════════════════
LAN Printing and Fax Products . . 9005
LANSpool for . . .
   LAN Manager/LAN Server Networks 9607
   NetWare 2.1X and 3.1X Networks . 9610
LANSpool FAX 9609
NetPort II Print Server 9600
NetPort II: What is it? 6014
NET SatisFAXtion 9590

Network Backup Servers
═══════════════════════════════════════
StorageExpress Backup Server . . . 6700
Network Backup Integrated Approach 6702

Copyright(C) Intel Corporation, 1992.

Figure 2-3 (continued): *Intel FaxBACK* **Cover Page and Typical Response**

TONE	MEANING
short high beep	over to you
low-high	ok, I heard that
long high-low	you goofed! (try again)
brief melody	you are at a starting point

Table 2-2: Good Use of Tones in Fax Response

This is in contrast to most voice applications, which rely heavily on use of verbal instructions to explain when to press what. Absent the beeps, one must wait for some time to pass in silence before deciding the recording is over. A short beep is immediately understood as "I'm done, now over to you"; when we hear the beep we know what to do.

In addition to handling documents by fax, *Intel FaxBACK* has the unusual ability to send software programs and data files to callers who have Intel SatisFAXtion 200, 400, or 400e fax modems in their computers. This unique feature is an attribute of the SatisFAXtion modems. Requesting data files or programs works the same way as ordering documents for fax delivery, except the result is a program or data file in your computer. This feature, if ever standardized in the computer or datacommunications industry, has incredible potential to expand fax response into "computer response." The result could be a new market and the growth the on-line industry has been wishing would happen for so many years.

Today PCEO estimates it handles in excess of a thousand calls a day, a volume that exceeds the calls taken by PCEO over the telephone. Intel figures this has saved the expense of ten full-time telephone operators. And Art King and associates now run a company that develops fax-on-demand systems. Not surprisingly, it's called FaxBACK Inc!

Io Publishing

Io Publishing is an innovative startup in the publishing business and one of the few companies that publishes a *daily* fax newspaper. *BioWorld Today* (see Figure 2-4) covers the fast changing biotechnology industry, where news of who is making what deals, filing what patents, or making new breakthroughs can literally move investments and markets overnight. The newspaper is read by nearly the entire biotechnology investment community and most of the senior management in the biotechnology business. Most depend on it as a primary source of news. *BioWorld Today* is a source of unique and timely information, and broadcast facsimile makes sure it gets there on time.

With reporters around the country, *BioWorld Today* gathers the news of the industry just as any daily newspaper would. Stories are written and edited much the same as they would be at a print daily. But that's where the similarity ends. At deadline time, work begins on the layout for the fax newsletter. Copyediting and layout are done in QuarkXpress, a desktop publishing program that runs on Apple Macintosh computers. All the news that's fit to fax is carefully placed on one, two, three, and occasionally more pages, on each page using a special template developed specifically for *BioWorld Today*.

This template determines the overall structure of the pages, specifying details such as type faces and graphics placement, and is optimized for use with fax. Specifically, it represents a finely tuned tradeoff between print quality, information density, and the need to economize on transmission time, the overwhelming component in delivery cost. Specific type faces were chosen that faxed well in standard resolution, and a layout was designed to make reading articles set in small type easy to read.

The details of the template were carefully matched to the process that turns a document in QuarkXpress into a fax document ready to transmit (see Figure 2-5). That process uses Ultrascript PC from PM Ware running on a generic AT-type 386 PC. PostScript type styles were used exclusively and were chosen to provide compatibility between the fonts used on the Macintosh by QuarkXpress and those used on the PC by Ultrascript PC.

TUESDAY
JULY 30, 1991

THE DAILY BIOTECHNOLOGY NEWSPAPER

VOL. 2, No. 147
PAGE 1 OF 2

Collaborative Gets Genome Grant

By Karen Bernstein
BioWorld Staff

Collaborative Research Inc. on Monday said it has received a $5 million, three-year grant from the National Center for Human Genome Research to test large-scale sequencing technology on the DNA of Mycobacteria.

The goal of the project is to test computer-assisted multiplex sequencing on large stretches of DNA. If successful, Bedford, Mass.-based CRI said it hopes the technology will halve the cost of DNA sequencing.

"The fantasy a few years ago was a cost of $1 per base pair," said Orrie Friedman, the company's founder, chairman and chief executive officer. "The target for this technology is 50 cents per base pair."

Originally developed by Dr. George Church and his associates at the Harvard Medical School, multiplex sequencing allows a researcher to read the results of multiple sequencing runs from a single electrophoresis gel. This greatly reduces the number of gels needed to sequence large amounts of DNA. Gels are then read automatically by computerized scanners. Church is one of several researchers who will collaborate on the project.

Sequencing the Mycobacteria genomes will allow CRI to target a medically important organism. Mycobacteria cause diseases such as leprosy and tuberculosis, which combined cause more deaths worldwide than any other bacterial disease. M. leprae affects about 10 million to 12 million people, mainly in Asia and Africa. M. tuberculosis infects 1.7 billion people worldwide.

Mycobacteria contain about 4 million base pairs in their DNA. The largest genome sequenced to date is that of the cytomegalovirus, which contains about 250,000 base pairs. CRI stock (NASDAQ:CRIC) closed up 13 cents on Monday at $1.25. ∎

Figure 2-4: *BioWorld Today* Daily Fax Newspaper

IAF BioChem to Test AIDS Drug

By Roberta Friedman, Ph.D.
Special to BioWorld

A nucleoside analog called 3TC will be entering clinical testing as an AIDS drug in August, IAF BioChem International Inc. announced Monday.

The tests, paid for by Glaxo Group Ltd. of London, will take place in several U.S. and Canadian medical centers. A similar trial will start simultaneously in Europe.

A European Phase I study in HIV-positive but asymptomatic patients ended in June, with results showing good safety and bioavailability, Francesco Bellini, IAF president and chief executive officer, told *BioWorld*.

The nucleoside analog is "different and specific" compared to the other analogs, AZT and ddI, now used or being tested as AIDS therapy. The structure of 3TC incorporates thiol in its sugar portion, Bellini said, and is a specific inhibitor of the viral enzyme reverse transcriptase.

Under a 1990 agreement, Glaxo is funding the development of 3TC worldwide and is also paying IAF $3 million Canadian (U.S. $2.6 million) a year for five years to continue basic research into nucleoside anti-virals. Glaxo is making milestone payments to IAF until 3TC reaches the market.

IAF has received $2 million in milestone payments and will receive another $1 million now, Bellini said. Once the drug is approved for sale, IAF will get royalties in sales outside of North America and royalties plus a 50 percent share of manufacturing profits with Glaxo within North America. IAF is based in Laval, Quebec.

Several patents have been filed on the analog molecule, and one is expected to issue soon in the United States, according to Jim McDonald, vice president of business development at IAF.

IAF stock (NASDAQ: BCHXF) closed at $24.25, up 50 cents, on Monday. ∎

Amgen's Neupogen Approved in Germany

Amgen Inc.'s Neupogen granulocyte colony stimulating factor has received marketing approval in Germany. Neupogen stimulates the production in bone marrow of white blood cells in patients receiving chemotherapy for certain cancers. It already has marketing approval in five other European countries and the United States.

Neupogen will be sold in Germany by affiliates of both Amgen and Roche Holding AG.

Stock of Thousand Oaks, Calif.-based Amgen (NASDAQ:AMGN) closed at $144.75, up $4, on Monday.

For BioWorld Customer Service, call 1-800-879-8790; in Japan, call COMLINE International Corp., 03-3486-0696.

BioWorld Today© is published every business day by Io Publishing Inc. President and Publisher, David Bunnell. Editorial and business offices: 217 South B Street, San Mateo CA 94401. Telephone (415) 696-6555; Fax (415) 696-6590; Washington Bureau (202) 662-7431. BioWorld and BioWorld Today are trademarks of Io Publishing Inc. Copyright © 1990 Io Publishing Inc. All Rights Reserved. NO PART OF THIS PUBLICATION MAY BE REPRODUCED WITHOUT THE WRITTEN CONSENT OF IO. To subscribe or to obtain photocopying rights, please call Subscriptions at 800-879-8790.

Figure 2-4 (continued): *BioWorld Today* **Daily Fax Newspaper**

When the newspaper is ready for a prepress check, a proof copy is generated the same way as a final copy. QuarkXpress writes to disk a PostScript file that contains the pages of the newspaper. That file is passed over a local area network (LAN) to the PC that runs Ultrascript PC. Ultrascript PC then processes the PostScript file into a set of TIF format files compatible with the Gammafax CP fax modems used in their FAXBLASTER multiline fax broadcasting system from DBC Associates. The FAXBLASTER then transmits the fax newspaper to a fax machine in the offices of Io Publishing, where it is proofread and approved for transmission.

The subscriber list is maintained using Filemaker on a Macintosh. At press time, this database is used to generate the list of subscribers and their fax numbers for processing by the FAXBLASTER and a service bureau. This list changes daily. The list of subscribers with fax numbers and the proofed TIF files containing the newspaper are then moved to the FAXBLASTER in preparation for transmission to domestic and European subscribers. A separate list and set of files are transferred to a service bureau for distribution in Japan.

Io Publishing uses AT&T Enhanced Facsimile Service as a backup in case of equipment failure or a local disaster such as an earthquake or a storm, and for extra capacity when the in-house equipment cannot provide the burst capacity needed to deliver special news issues that occasionally require immediate daytime delivery with a much shorter than normal lead time.

BioWorld Today is transmitted to arrive at subscriber fax machines before the start of each business day. (Publishing deadlines are Sunday to Thursday nights.) It has a circulation of approximately a thousand readers mostly across the United States with only a few internationally in Europe and Japan. Transmission begins at 2300 hours Pacific time (Io Publishing is in San Mateo, California) and all transmissions are complete by 0800 hours.

Io Publishing got started producing *BioWorld Today* as a by-product to an on-line information service the company was developing. The original concept for *BioWorld Today* was to produce a promotional device that would give

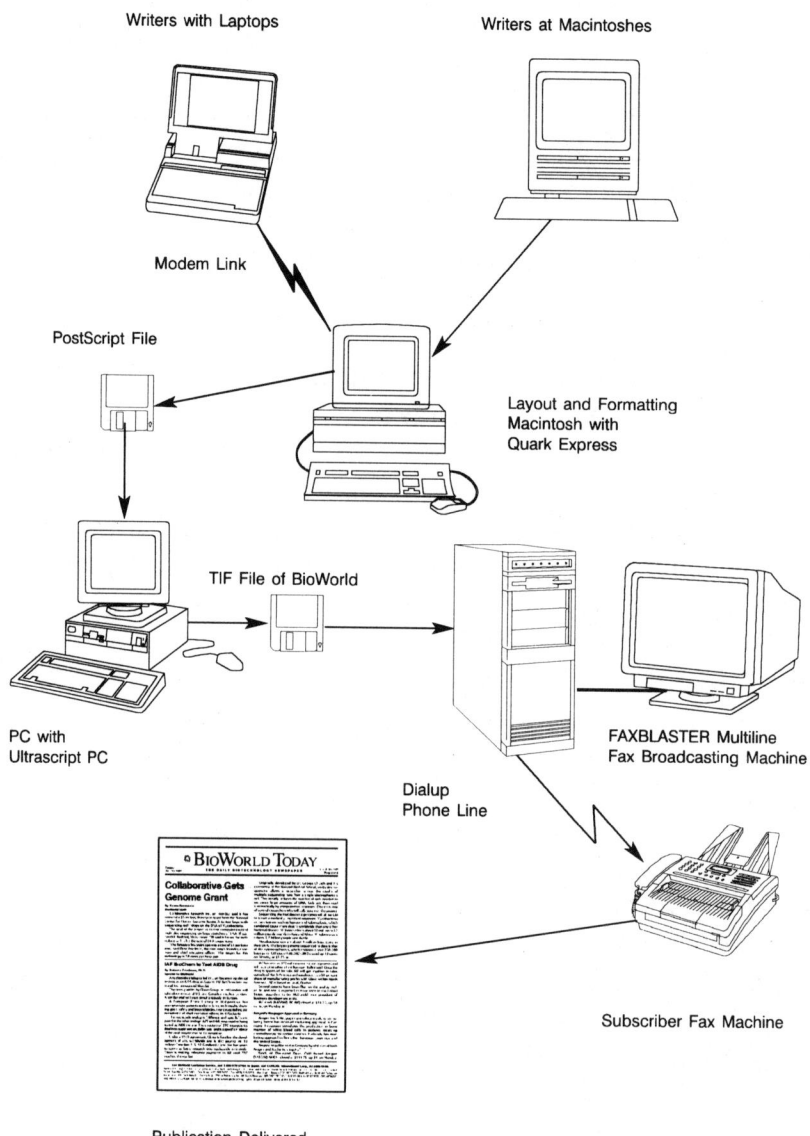

Writers with Laptops

Writers at Macintoshes

Modem Link

PostScript File

Layout and Formatting
Macintosh with
Quark Express

TIF File of BioWorld

PC with
Ultrascript PC

FAXBLASTER Multiline
Fax Broadcasting Machine

Dialup
Phone Line

Subscriber Fax Machine

Publication Delivered

Figure 2-5: *BioWorld Today* **Fax Publishing Process**

prospective on-line users a sample of what they could get. In the end, the convenience and timeliness of getting information by fax proved to be overwhelming. Readers of the fax paper signed up faster than on-line users. *BioWorld Today* is now the mainstay of Io Publishing in the biotechnology area.

Borland International Tech Fax

Borland International is a large maker of software products, including databases and programming languages. A major challenge faced by the company is meeting the need for customer support. Users of software products often require detailed technical information as well as news of product changes and new product introductions. The company's *Tech Fax* service provides product, technical, and company information on-demand and around the clock. It is a good example of how to set up a fax-on-demand system for maximum ease of use (see Figure 2-6).

Callers reach *Tech Fax* by calling 800-822-4269 from any touch-tone telephone. A brief recorded announcement identifies the service and offers callers only three choices:

1) Hear a recorded message describing *Tech Fax* and how to use it.

2) Order a document by number.

3) Receive one or more catalogs that identify documents by number.

To: 4157760615   From: TechFax   1-24-93 6:18pm   p. 1 of 10

B O R L A N D

P.O. Box 660001. Scotts Valley, CA 95067   TechFax: 1(800)822-4269

To: 4157760615   Date: 1-24-93

From: TechFax   Page 1 of 10

```
                 Try Borland's other great support services!
                   Available 24 hours/day - 7 days/week!

TechFax      Our automated fax retrieval service will send free technical
             information to your fax machine.  To receive a catalog of
             available documents and information on using this service call
             1-800-822-4269

On-Line      Use your modem to download technical information, applications,
             and sample files.  The Borland Download BBS can be reached by
             dialing (408) 439-9096.
             To access our forums on CompuServe, type GO BORLAND
             on BIX, type JOIN BORLAND and on GEnie, type BORLAND.

800-         This fully automated voice response system provides up-to-date
Automated    technical information.  Support for Quattro Pro, dBASE IV,
Support      and Paradox is available by calling 1-800-524-8420
```

Figure 2-6: Product Support via Fax-on-demand

This initial structure is clear and concise. Restricting the number of possible choices to three makes it virtually impossible for a caller to get lost in the complicated maze of voice menus typical of many such systems. The recorded message about the service is short and describes the purpose and operation of *Tech Fax* in plain language. The ordering of documents by number allows callers to enter requests for multiple documents at a time without cycling through repeated voice prompts. Order entry is completed with the entry of the telephone number of the destination fax machine, and transmission takes place on a return call.

For callers who need lists of documents, the third choice offered is to receive one or more catalogs that list the documents by topic (see Figure 2-7). This option has been quite carefully crafted. Documents are listed in catalogs based on subject or product area. For example, all documents about the database product Paradox are listed together. Altogether there are thirteen catalogs, but the voice recording for catalogs gives options for only the top four, plus a fifth choice, which is a listing of all available catalogs.

```
Main Menu
        1: Instructions for using Tech Fax
        2: Document ordering
        3: Catalog ordering
                1: Databases
                2: Applications
                3: Languages
                4: News
                5: Catalog of all catalogs available
```

Figure 2-7: Structure of Borland *Tech Fax*

This structuring has beneficial effects. It responds to the most likely needs in the quickest fashion by putting the most often requested information in the three choices of catalogs available directly from the voice menu. At the same time it preserves ready access to a far greater—although less often used—body of information while avoiding long-winded voice prompts, by using fax instead of voice to deliver a much larger list of catalog choices.

Tech Fax is a no-nonsense implementation, and it seems to be quite heavily used (see Figure 2-8). While testing the system for this book, a backlog of outgoing faxes seemed to develop during the daytime, indicating it was getting a high volume of requests. Fax responses could be delayed during the day for up to an hour. These delays were not present at night or on weekends.   From a design standpoint it is interesting to note that the only graphics that appeared in the tests conducted for this book were on the cover pages sent out by *Tech Fax*. Nearly all other documents were digital direct faxes of ASCII text.

Perhaps the only suggestion that could improve *Tech Fax* would be to switch from sending in fine mode to standard mode. This would cut in half the transmission times, effectively doubling throughput and halving transmission costs. And nearly all documents have no graphics, only ASCII text converted to fax, there would be no difference at all in document appearance. You can see this for yourself by trying both *Tech Fax* and the previously mentioned Intel FaxBACK. The former uses fine mode and the latter uses standard mode, but the text should have the same quality. Lastly, the cover sheet, shown in Figure 2-8 contains some very costly screening in the *Tech Fax* logo; this could be redesigned to greatly reduce the transmission time of the cover page, which goes out with every return call. The *content* of this service, however, is exactly the kind of information callers need and want—and that is clearly a requirement for a successful fax-on-demand application.

To: 4157760615 From: TechFax 1-24-93 6:31pm p. 2 of 7

PRODUCT	:	Paradox	NUMBER	:	1149
VERSION	:	4.0			
OS	:	DOS			
DATE	:	December 2, 1992	PAGE	:	1/6
TITLE	:	Upgrading to Paradox 4.0			

Intended Audience:
All Paradox users upgrading to Paradox 4.0. Aimed especially at users of Paradox 3.5.

Prerequisites:
Should be familiar with a version of Paradox prior to 4.0.

Purpose:
This Technical Information sheet provides a checklist of tips on upgrading Paradox 3.5 to 4.0. It also includes some tips on upgrading applications to take advantage of the new features. Most of the information applies to 3.0, and users of 2.0 should still find this useful.

There are many other Technical Information sheets containing information about specific aspects of using Paradox 4.0, and the number is growing. It would be a good idea to download our catalog on a regular basis. Here are some that are current:

#1063 An Explanation of 4.0 Table Locks
#1107 Manipulating Window Frames
#1109 Record Locks in 4.0
#1110 Converting PPROG Applications to 4.0
#1111 Summary of features and benefits of Paradox 4.0
#1112 Using IIF() in Reports
#1123 Introduction to DYNARRAY Branching

USER INTERFACE & GENERAL ISSUES

- o Paradox 4.0 defaults to the new interface. This user interface looks and behaves differently from that found in Paradox 3.5. It may initially be more comfortable to work with the Compatible User Interface. To switch to Compatible Mode, start Paradox with the command-line switch -COMPMODE, or press <ALT-Space> ¦ Interface ¦ Yes within Paradox to switch between modes. You can also make this a permanent default by doing <ALT-Space> ¦ Utilities ¦ Custom ¦ Standard Settings ¦ Interface ¦ Compatible.

- o Compatible Mode contains most of the feature improvements in Paradox 4.0, including memo fields, improved secondary indexes, and the new locking mechanism.

Figure 2-8: A Typical No-Nonsense Tech Fax

The Foresight Institute

The Foresight Institute is a membership organization formed to promote the development and sharing of knowledge in the esoteric field of nanotechnology. (Call the institute's fax-on-demand system, at 415-948-8310, to find out more about nanotechnology.)   Like many member organizations, the institute has limited resources and is always seeking ways to leverage the time of staff and volunteers, while staying responsive to members and potential members.   One of the institute's primary tasks is to dispense information on nanotechnology and institute activities to both members and prospects.

The Foresight Institute publishes a monthly member newsletter as its main form of communication with its approximately 1,200 members.   The newsletter, which is mailed to members, describes upcoming events and news in the field, but it cannot possibly contain all the information that members are likely to want, since a wide variety of articles, papers, and other documents are new each month.   As a service to its members and to promote interest in nanotechnology, the institute will send out copies of certain materials by mail, but this is both a burden on the organization and an extra expense.

In the past, to request a document, members called a phone number and left a voice message describing the document they wanted and where to send it.   A staff person would then transcribe these recorded voice messages and then manually fulfill the requests.   Chris Peterson, the editor of the newsletter, conservatively estimates that fulfilling a typical request for a document costs $5.25, as shown in Figure 2-9.

To help reduce the costs involved in handling these requests and increase availability of the documents, the Foresight Institute started using the MessagePost, a very low-cost fax-on-demand system from DBC Associates. Documents previously copied and sent by mail were transmitted into the MessagePost using an ordinary fax machine.   A small "library" was created and a list describing the documents available was also stored in the MessagePost system.

Postage	$1.00
Copying	$1.00
Envelope	$0.25
Labor (10 minutes)	$3.00
TOTAL COST	**$5.25**

Figure 2-9: Fulfillment Cost Per Document Requested

Now, both members and prospective members can retrieve documents from a compact library stored in the institute's MessagePost. Operation is simple. People who want a document call the fax-on-demand system using the handset on their fax machine. Callers who are already familiar with its operation simply follow a very brief set of voice prompts to indicate which documents they want. They then engage their fax machine by pressing the START/COPY or SEND/RECEIVE button, and immediately begin reception (see Figure 2-10). Documents are delivered while they wait.

For new callers who are unfamiliar with how to operate the system, the procedure is even shorter. They just call from the handset on their fax and press the START/COPY or SEND/RECEIVE button. Simple written operating instructions and a list of available documents are then sent to them by fax. This method avoids long voice prompts and extensive menus which can be frustrating and confusing for the novice caller. It also allows the *Foresight Institute* to list a large number of documents along with their descriptions, without having to use voice prompts. It takes maximum advantage of the strength of fax-on-demand by sending lists and descriptions of documents and instructions by fax and taking orders using voice recordings and touch-tone.

Loading documents into the fax response system is equally easy and straightforward (see Figure 2-11). Using a touch-tone telephone or touch-tone handset attached to the fax machine, a system operator calls up and presses touch-tone buttons in response to prerecorded voice prompts. The operator uses

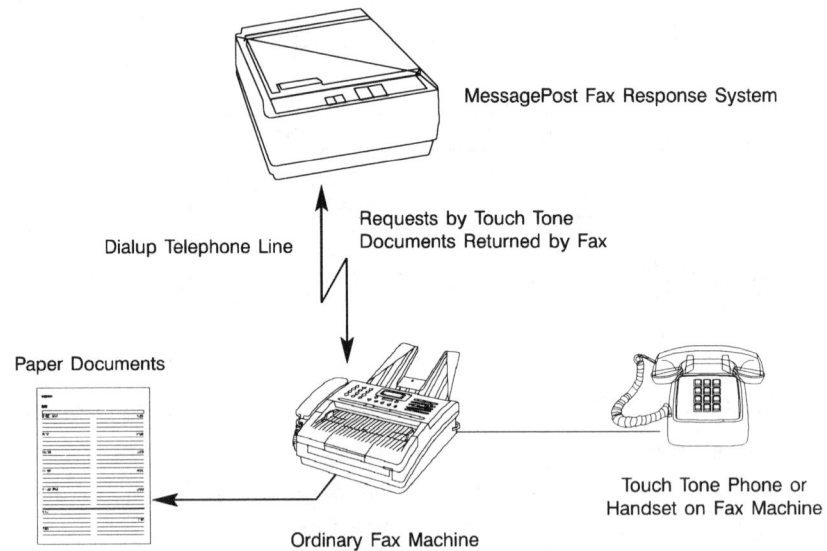

MessagePost Fax Response System

Dialup Telephone Line

Requests by Touch Tone
Documents Returned by Fax

Paper Documents

Touch Tone Phone or
Handset on Fax Machine

Ordinary Fax Machine

Figure 2-10: Getting A Document

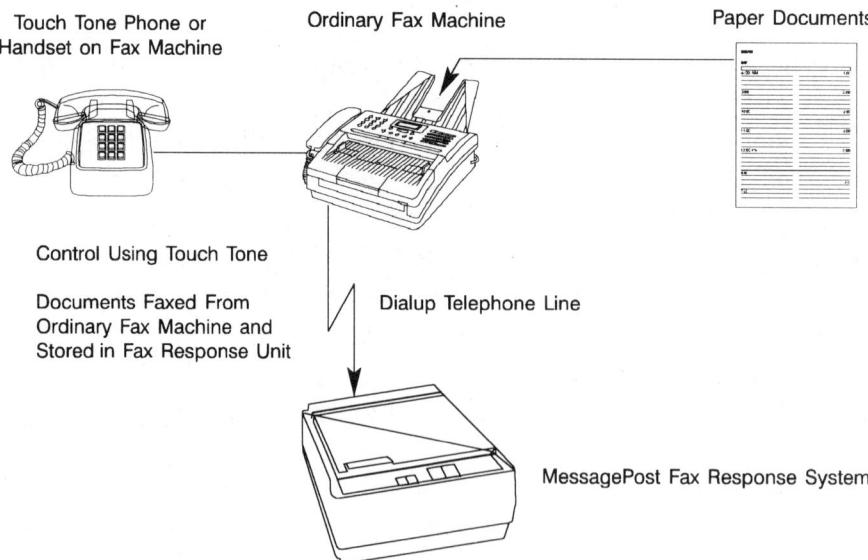

Touch Tone Phone or
Handset on Fax Machine

Ordinary Fax Machine

Paper Documents

Control Using Touch Tone

Documents Faxed From
Ordinary Fax Machine and
Stored in Fax Response Unit

Dialup Telephone Line

MessagePost Fax Response System

Figure 2-11: Loading A Document

the fax machine in manual mode to upload a new document. Functions are provided for adding, removing, and changing the numbers of documents.

For the Foresight Institute this "electronic literature bin" has reduced by thousands of dollars the cost of processing document requests, and has greatly lowered the space required for carrying an inventory of documents. Members benefit by having instant access to information they want. Peterson says this is a great way for small membership organizations, particularly those that are nonprofit, to hold down costs while increasing available services.

Cruising World Magazine

Cruising World is an upscale boating magazine with a circulation of 135,000. The management had a problem most publishers would envy. Their ten-year-old reader service, *Another Opinion,* had become exceedingly popular and was overloading the staff (see Figure 2-12).

Another Opinion matches readers who have questions on specific boats with people who own boats of the make and model in which the reader is interested. Prior to automation using on-demand fax, this job was handled by the magazine staff as a free service. To make requests, readers would write a letter to the magazine. A staff member would then consult a database of approximately 2,000 boat owners and generate a printout. A few weeks later the results would arrive by mail.

As the number of requests per month built to 1,800 (that's 90 per working day!) the burden became overwhelming, and *Cruising World* saw an opportunity to turn a loss leader into a new revenue source. The magazine teamed up with Instant Information of Boston, Massachusetts, a fax-on-demand service bureau.

Now, readers who want to make requests for *Another Opinion* are first directed to an index of boat types in the magazine (see Figure 2-13). Each boat type in the database is identified by a numeric "boat ID." Instant Information now

Another Opinion

5 John Clarke Road
Newport, RI 02840
401-847-1588

REQUESTED INFORMATION.   PLEASE HOLD AT THIS FAX MACHINE.

Dear Reader,

Thank you for calling CRUISING WORLD's "Another Opinion." Below you
will find the requested names and addresses of up to five people who
have agreed to share their personal opinions or experiences about the
vessels. Please remember to enclose a self-addressed, stamped envelope
when writing to these contacts.

```
BOAT ID# 1013      BOAT NAME: WESTERLY NOMAD             BOAT LENGTH:
------------------------------------------------------------------------
Neal  Dodge                       H.(802)864-9852   W.(802)241-2885
169 Home Avenue, Burlington, VT 05401                   6/17/1987
------------------------------------------------------------------------
Brenda  Cooper                    H.(916)422-8234   W.(916)324-6439
6005 Gloria Drive,#19Sacramento, CA 95822               4/24/1990
------------------------------------------------------------------------
Dave  Reynolds                    H.(603)964-6446
Post Office Box 66, Rye, NH 03870                        7/02/1987
------------------------------------------------------------------------
Cornelius C Kerkstra
7083 West Wood Drive, Jenison MI 49428                  10/20/1987
------------------------------------------------------------------------
Stephen J Romanoff                H.(207)781-5209
304 Foreside Road, Falmouth ME 04105                    1/01/1989
------------------------------------------------------------------------
```

Again, thank you for calling. We hope that the requested information
will be helpful. Should you have any questions, please call Elise
Black at CRUISING WORLD at 401-847-1588.

Fax: 401-848-5048
Telex: 467953

A Publication of The New York Times Company

Figure 2-12: Cruising World "Another Opinion"

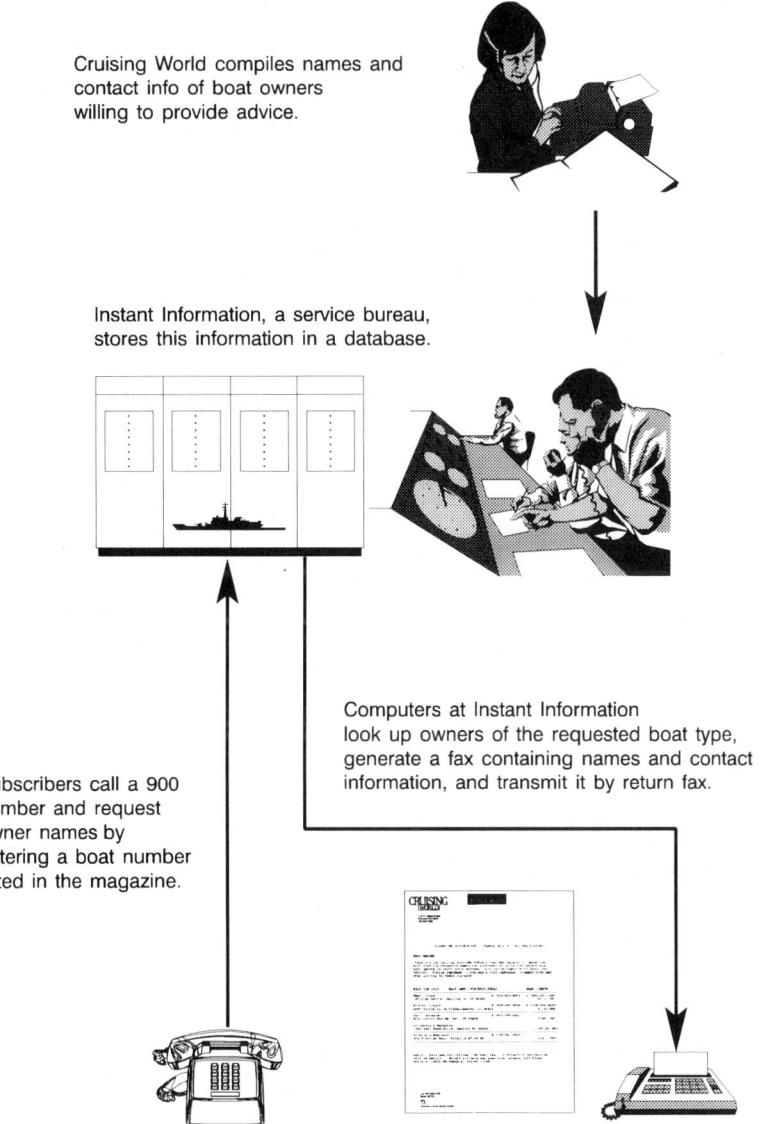

Cruising World compiles names and contact info of boat owners willing to provide advice.

Instant Information, a service bureau, stores this information in a database.

Computers at Instant Information look up owners of the requested boat type, generate a fax containing names and contact information, and transmit it by return fax.

Subscribers call a 900 number and request owner names by entering a boat number listed in the magazine.

Figure 2-13: Cruising World Functional Diagram

maintains the database of boat owners and ID numbers using a personal computer and a customized version of Paradox from Borland International. The reader who wants *Another Opinion* looks up the numeric code for the boat of interest and, using a touch-tone telephone, calls a 900 pay-per-call number. This number goes to a fax-on-demand system at *Instant Information*. The fax-on-demand system plays recorded instructions, and the reader is told how to enter the boat ID using the touch-tone buttons on the telephone. The ID is routed to a database of approximately 2,500 owners. The database then selects names based on the boat ID and generates a report containing the names, addresses, and telephone numbers of up to five owners. This report is converted to fax by a UNIX-based system, which composes the document from three parts: a header which is the *Cruising World* letterhead, the report text, and a footer, which is the bottom part of the letterhead. The final fax document is transmitted as shown using Brooktrout TR112 fax boards in a computer. Readers without touch-tone phones can leave a voice message to make the request.

Cruising World charges $3.95 per call, which is billed to the caller's telephone account by the 900 number. The staff spends less time handling requests and readers get their information in minutes instead of weeks. And fax-on-demand has created another all-around winning situation.

U.S. Navy/NOAA Joint Ice Center

The *U.S. Navy* in joint operations with the National Oceanic and Atmospheric Administration (NOAA), supports a service that provides up-to-date charts that show where large chunks of ice are floating in the world's oceans. These maps help mariners stay clear of areas containing icebergs and other floating hazards to navigation. This service has a distinctly military, no-nonsense style to its introduction. The list of available maps is equally brief and to the point (see Figure 2-14).

```
12/22/92 22:38:27;     (381) 763-3198->              ; NAVY/NOAA JIC    Page  1
MAR-11-92 WED 18:48 NAVPOLAROCEANCEN SUIT.MD          P.82
```

ATTENTION***ATTENTION***ATTENTION***ATTENTION***ATTENTION

NAVY/NOAA JIC DIAL-IN USER'S GUIDE

INTRODUCTION: The Navy/NOAA Joint Ice Center has instituted dial-in facsimile distribution capabilities for all standard ice analysis/forecast graphic products. Dial-in users will be prompted by a voice menu allowing product selection and reception using a touch-tone key pad and a facsimile machine. The purpose of this User's Guide is to provide dial-in instructions and a user's list of products available.
For a backlog of charts that may have been missed during this transitional period please contact:

CUSTOMER SERVICE BRANCH
NATIONAL CLIMATIC DATA CENTER
FEDERAL BLDG.
ASHEVILLE, N.C. USA
28801-2696

Or call Mr. Sam M. McCowen at (704) 259-0272.

EQUIPMENT: The JIC facsimile auto-polling system is compatible with CCITT Group 3, CCITT Group 2 and North American Group 1 type facsimile machines.

PROCEDURE: Dial the Joint Ice Center on either (301) 763-3190 or (301) 763-3191 [763-3190/3191 on FTS] and follow the voice prompt instructions for product selection from a TOUCH-TONE keypad.

Once you have entered your product selection wait for an additional voice prompt directing the system to the product location, make the appropriate selection and wait for the facsimile tone to sound. Then press the start or receive button on your fax machine to down-load the graphic product.

The fax link will be terminated following product transmission, any further transactions will require a separate call. It is recommended that first time callers press numeral "1" to receive a products list and instructions to use a reference for future transactions. [NOTE: The products list and instructions are identical to this page].

Figure 2-14: Joint Ice Center Instructions and Product List

```
12/22/92 22:39:26;      (301) 763-3190->              ; NAVY/NOAA JIC      Page  2
MAR-11-92 WED 10:41 NAVPOLAROCEANCEN SUIT.MD          P.03
```

PRODUCT LIST: The following is a list of standard ice products
                available through touch-tone key pad selection.

```
        MENU
KEYPAD #        PRODUCT

   1            INSTRUCTION/PRODUCT LIST

   2            ARCTIC ICE ANALYSIS/FCST

                    SUB-MENU
                    KEYPAD #    PRODUCT

                        1       NORTHERN HEMISPHERE ANALYSIS
                        2       EAST ARCTIC ANALYSIS
                        3       WEST ARCTIC ANALYSIS
                        4       NORTHERN HEMISPHERE ANALYSIS & 30 DAY FCST

   3            ANTARCTIC ICE ANALYSIS
                        1       FULL ANTARCTIC ANALYSIS AND  96 HR  ICE
                                FORECAST FOR WEDDELL SEA ONLY
                                (THURSDAY - MONDAY AFTER 2100 UTC)
                        2       WEDDELL SEA EDGE UPDATE AND 72 HR  ICE
                                FORECAST
                                (MONDAY - THURSDAY AFTER 2100 UTC)
                        3       SEASONAL OUTLOOK FOR WESTERN ROSS SEA

   4            ALASKA REGIONAL ANALYSIS

   5            GREAT LAKES

                    SUB-MENU
                    KEYPAD #    PRODUCT

                        1       ANALYSIS
                        2       30 DAY FCST
                        3       ST. MARY'S RIVER
```

PROBLEMS WITH FACSIMILE TRANSMISSION MAY BE ADDRESSED VIA
PHONE (301) 763-7310, FAX (301) 763-4621, OMNET MAIL BOX
"NAVY.NOAA.JIC", OR TELEX 7402918 ANS:NOAA.

Figure 2-14 (continued) : Joint Ice Center Instructions and Product List

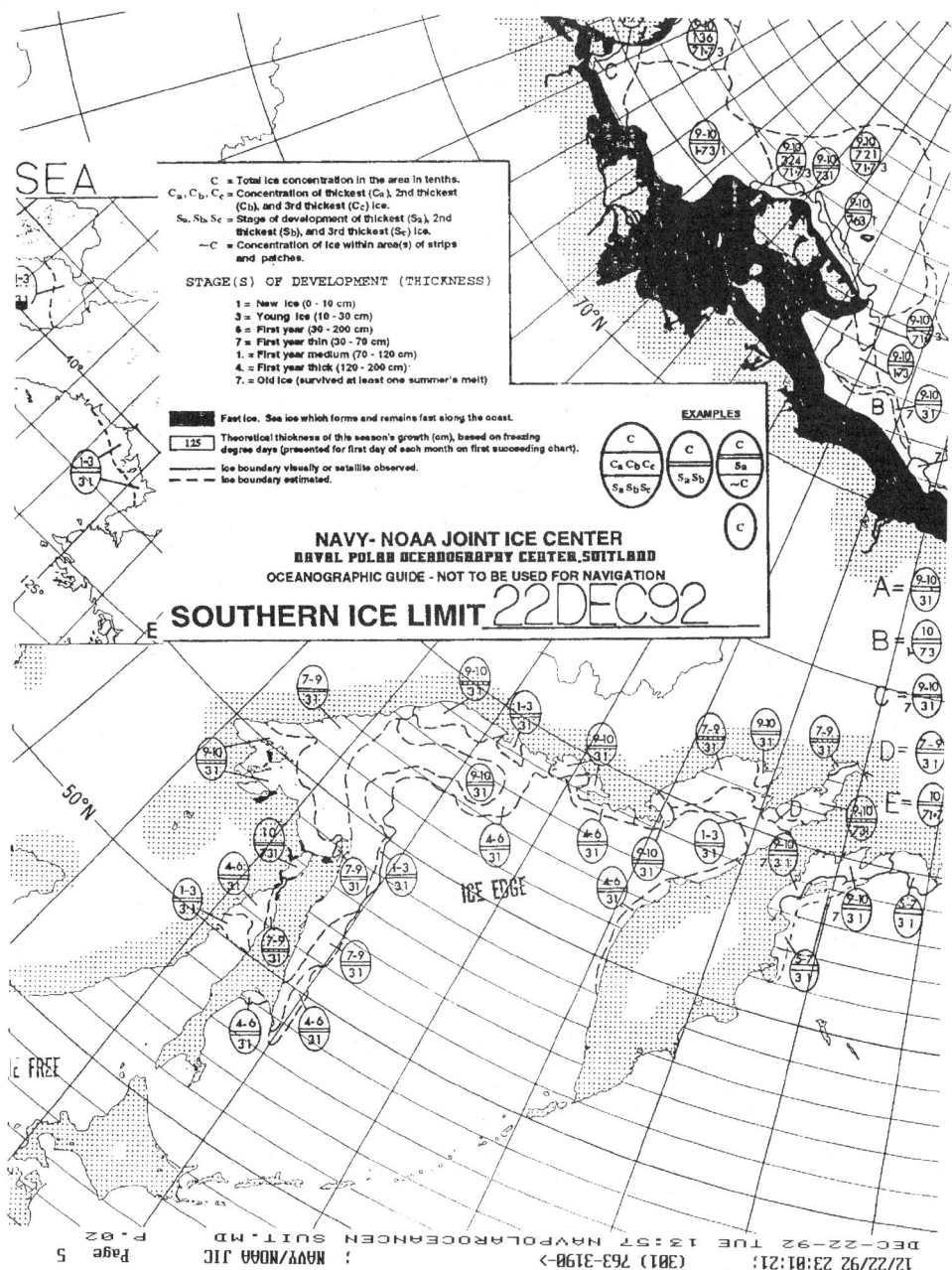

Figure 2-15: A Typical Ice Map

Mariners and others who wish to obtain ice analysis information can call 301-763-3190 from their fax machines and select documents by using touch tones. The entire operation is one-call, so callers must call from their fax machines. The Ice Center does this to avoid paying for potentially expensive return calls. More than 140 requests a week come in to the center's fax-on-demand system, which consists of a Zenith-286 computer, fax boards, and fax-on-demand software, all of which were purchased off the shelf and pieced together by the computer department.

Ice analysis maps, (see Figure 2-15) are hand drawn on large pieces of paper and then reduced to letter size using a reducing photocopier. The letter-size versions of the maps are then transmitted into the fax response system using an ordinary fax machine. The reduction and loading into the fax system is the only routine work required to support this fax-on-demand service.

Prior to installing the fax-on-demand system, the Ice Center mailed out ice maps to over one hundred subscribers each week. This was time consuming and costly and required the time of three to four people for one full day a week to assemble the customized map packages. Replacing this more traditional publishing method with fax-on-demand publishing has resulted in savings to the Ice Center and faster service to customers. Today the Ice Center estimates it has approximately 140 regular callers who use the fax response system to get ice maps. The increased availability and convenience of receiving this information by fax is a likely cause for the increase in usage.

Ottawa Police Crime FAX

CrimeFAX was started to help stop credit card fraud in the banking system by allowing the Ottawa Police to broadcast crime alerts to banks. It was funded by the Canadian Bankers Association and estimates to date say it has already more than recovered its cost in the crimes and associated losses it has prevented.

This unique application required a way for the Ottawa Police to quickly compose a document from existing paper sources and deliver it to banks within a period of no more than a few hours. A unique setup involving a computer-controlled multifunction fax machine was used. It provides for rapid document creation, proofing, and transmission with a minimum of work done by law enforcement staff. Total cost of the setup was about $7,500 Canadian.

When trouble starts, a *CrimeFAX* bulletin is literally cut and pasted together from various paper sources available to the Ottawa Police. They use this method because it is much faster than trying to compose and format a document using computers or desktop publishing tools. While it may not be as pretty as a more formally produced document, speed is considered more important in preventing large banking fraud crimes.

The cut and pasted *CrimeFAX* hard copy is then scanned into the memory of a Canon L770 fax machine which is controlled by a PC running SciFAX from Cognition Sciences Corporation. The Canon fax machine has enough memory to store the entire *CrimeFAX* bulletin. The PC running SciFAX is used to control the Canon L770 fax machine because it is simpler and faster to use the on-screen menus of the computer than the special function buttons on the fax machine. The PC running SciFAX stores the names and fax numbers of all the banks that need to receive *CrimeFAX*. It also directly operates the fax to control scanning, storing, and broadcasting by the machine.

The entire process is faster and easier to use than either fax machine alone or DTP process and a fax modem. Scanning takes about five seconds per page, and the transmission of *CrimeFAX* to about two hundred banks takes place without any human intervention. Once the document is scanned into the fax and the broadcast is started, the process is automatic.

More Ideas

Here is a summary of ideas which may help you think of how you might use fax power in your organization.

- A police department uses an automatic fax broadcast relay to speed incident reports from officers in field offices to a dozen critical locations.

- Grass-roots political organizations use fax broadcasting to send out alert and action notices to members. Telephone trees are eliminated.

- A magazine uses fax-on-demand to provide reprints of articles and reader service.

- Once a month a reseller of electronic components uses fax broadcasting to send updated product price and availability information to its entire database of customers.

- A catalog merchant uses fax-on-demand to provide access to detailed product information, including instructions, manuals, and manufacturers' warranties.

- A major metropolitan newspaper uses a low-cost fax-on-demand system to distribute sports news and scores that are in high demand but too specialized to print in the paper.

- Restaurants use a simplified fax-on-demand to dispense a single document: the menu or menu of the day.

- Real estate listings are available via fax-on-demand. A simple set of voice prompts allows callers to select properties by price, location, and major features such as number of rooms. The fax response system drives a computer database to obtain its answers.

- A fax-on-demand system is used as a literature dispensary by the sales force of a company. Instead of faxing materials directly, the sales force calls the fax response system and tells it to send the information. Customers and prospects may also use it directly.

- A stockbroker uses a database and multiple fax modems in a personal computer to send news of hot investment opportunities to clients. The broker can reach more than three hundred clients in an hour using four lines.

- Banks start offering statement information using on-demand fax. Customers can call in and request a current statement at any time. For security, statements are transmitted only to a predetermined number, addressed to the account holder.

- Travel agents receive fax broadcast alerts from airlines and other travel organizations announcing overnight price changes, specials, and other promotional information.

CHAPTER 3

FAX BROADCASTING

There are two basic methods for delivering documents by fax: *broadcast* and *on-demand*. Broadcast fax is most often used to send the same document to many recipients, similar to a "mass mailing." Fax-on-demand is mostly used to handle repeat requests for differing information, much like a literature rack or automated teller. In this chapter we will describe broadcast fax, and in Chapter 3 we will cover fax-on-demand. We will see how broadcast fax works and discuss how to use the different tools available for broadcasting. Whatever your interest, it is probably useful for all readers to read the section on conventional fax machines.

Broadcast Facsimile

Broadcast facsimile is transmission of a document, by fax, to multiple recipients. For each recipient, a separate call is made from the sending fax machine or broadcast system to each receiving machine. During this call the document and any other material, such as a cover page, is sent. Although it is called "broadcast" because the same fax is sent to multiple recipients, fax broadcasting is really fax *narrowcasting*. And that is part of the benefit of broadcast fax. You the sender determine exactly who gets your document and when. Figure 3-1 shows a diagram of broadcast fax.

SOURCES OF FAX DOCUMENTS FROM
DISKETTES OR MACHINES

DESTINATION FAX MACHINES

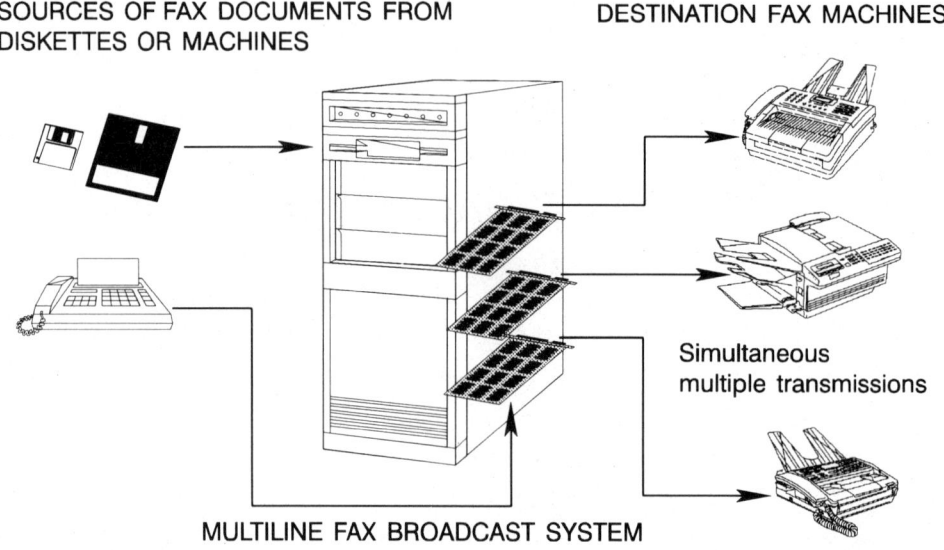

MULTILINE FAX BROADCAST SYSTEM

Simultaneous
multiple transmissions

Figure 3-1: Broadcast Fax

Using Conventional Fax Machines

There are several ways to broadcast faxes. The most readily available is the regular office fax machine. It's already there, you probably already know how to work it, and it may have some useful features to help you with a broadcast task. An office fax machine is suitable for broadcasting to smaller numbers of people, up to five or six. However, for larger numbers of transmissions, using a regular fax machine becomes impractical. It is too time-consuming and unreliable to have an operator stand over a machine to make sure all pages of a document feed correctly. And fax machines handle only one transmission at a time, further limiting transmission capacity. Later we will discuss other methods, such as using computer-driven fax boards, that get around these limitations. In this section we will look at how to use fax machines and some of their features that are useful in handling smaller numbers of transmissions.

Even the most basic fax machine has a scanner (into which the transmitted documents are placed), a printer, and a keypad for dialing. Better machines have additional features such as document feeders, handsets, sheet cutters, memory, and stored number dialing. A few even have advanced "relay" features, which allow them to receive a fax and resend it automatically to a predetermined list.

If you are only going to send a small document to five or even ten people, a basic fax machine with handwritten cover sheets or the new stick-on notes that contain cover page information will probably get the job done without too much fuss or expense. If you are sending to personal fax machines where the recipient will receive your transmission directly, you can save time by eliminating the cover page. While it is true a person has to stand by and feed all the pages of a document through the machine for each transmission, if the amount of transmitting is small, this is the easiest and probably fastest way to "just get the job done" with what you are most likely to have on hand.

Stored Number Dialing

If your demands go beyond a handful of transmissions, using higher-powered fax machines with more features begins to make sense. The first feature to look for is *stored number dialing*, often marketed under other names, such as *rep (repertory) dialing, one-touch numbers* or *number memory*. Stored number dialing can save you the time and hassle of looking up and dialing the fax number of each recipient. A machine with stored number dialing will dial a preprogrammed number for each stored number button it has. Before buying a machine with this feature you should carefully consider how many stored numbers you are likely to need and get a machine with sufficient capacity. Also, make sure the numbers can be easily programmed. Some machines let you enter in alphanumeric identification of each stored number. This is convenient—more so when the machine provides for convenient printing of a directory showing what numbers are programmed for which buttons. This is a try-before-you-buy feature.

Sheet Feeder

There are few machines today that do not come with a sheet feeder, but some machines have limited or less reliable sheet-feeding capacity. A sheet feeder can greatly ease the burden of transmission by allowing an operator to perform other functions while the machine handles the pages. When looking for a machine to meet your transmission needs, be sure it has a sheet feeder that can handle the number of pages in your documents.

Memory Machines

A fax machine with *memory* can do even more work for you. Provided your document does not exceed the memory capacity of the machine, you can feed your document into the machine to store it before transmission. Once the document is stored, you then instruct the machine to dial one or more stored numbers. Most memory machines have a storage limit of ten to twenty-five pages which is more than enough for most documents one would want to fax. The exact number of pages that will fit in memory depends on the content of each page. The section on performance issues in Chapter 3 covers this in detail. Memory feature saves the work of an operator who would otherwise need to stand by the machine and feed in the document once for each transmission. Note, however, that unless you use the same cover sheet for all transmissions, there is no way to send out personalized cover sheets containing the name of each recipient. This is one of the biggest drawbacks of broadcasting with regular fax machines versus computer-driven fax boards. As we will see later, using a computer-driven fax board can solve the problem of addressing cover sheets quite easily.

Delayed Transmission

Some memory machines can memorize numbers as lists and store documents in memory for delayed transmission. This allows you to save on transmission costs by delaying fax transmissions for when calling rates are lower. You might use such a machine in this way: First, set up the machine for transmission by storing all the numbers and then scanning the document into memory. Then, program a list into the machine. Finally, tell the machine to begin sending the stored document to the list starting at a specific time of day, such as 11 pm. By waiting until then to send, you lower your transmission costs.

Transmission Reports

Another feature to consider when using a fax machine for broadcasting is the transmission report. Getting a transmission report is very important because if you are using an unattended machine to automatically send transmissions, you will need to know whether the transmissions you tried to send were received. Look for a transmission report feature that is easy to read and contains enough information so you can figure out why a transmission may have failed. Most machines print transmission reports the same way they would print a received fax.

Relay Fax Machines

If your transmission needs increase beyond that of a single fax machine you may want to consider a network of *relay* fax machines (see Figure 3-2). A relay machine will receive a transmission from another, compatible relay machine and then resend the transmission to a preprogrammed list of numbers. The relay machines do not need to be specially connected but they do need to be compatible—usually machines made by the same manufacturer are. The machine that does the retransmissions need not be located in the same place as the sending machine. This kind of setup is most beneficial when you have a well-established set of recipients, or when they are clustered in certain areas that allow economy of transmission. A well-placed relay machine can even fully automate the process of routing information around a company.

Successful use of a relay depends somewhat on having a stable broadcast list. If the list of recipients changes with each transmission, then reprogramming a remote relay machine would be impractical, since the whole point of having a relay machine is to automatically retransmit a received fax to a somewhat constant list.

Another benefit is that relays can help you save on long distance transmission costs, particularly for international transmissions. If the recipients are clustered in a "local" calling area then you can send the initial transmission a long distance and have the relay resend it at local rates.

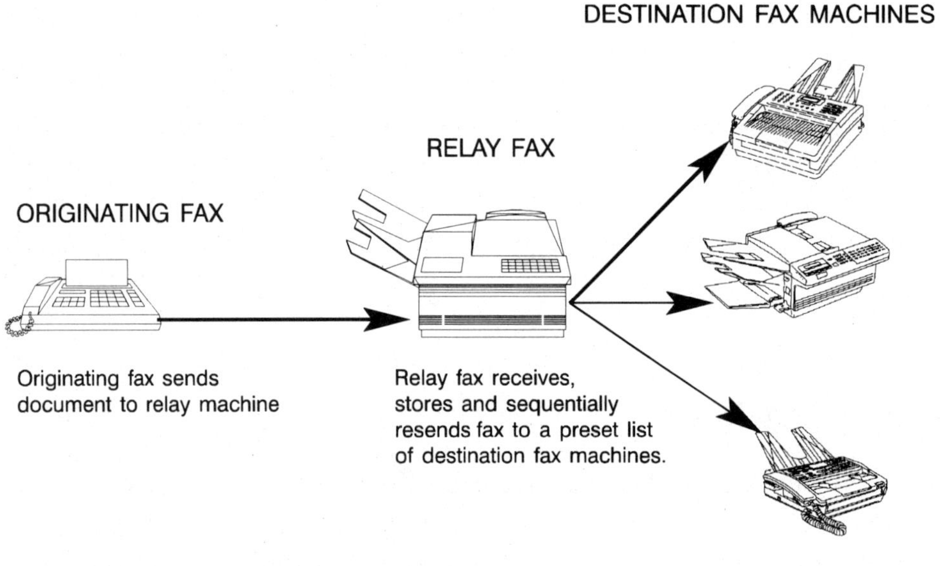

DESTINATION FAX MACHINES

RELAY FAX

ORIGINATING FAX

Originating fax sends
document to relay machine

Relay fax receives,
stores and sequentially
resends fax to a preset list
of destination fax machines.

Figure 3-2: Relay Fax Operation

Take the following examples. Company X wants to send internal bulletins from its headquarters in the U.S. to five of its sales offices located in England. Company X would place a relay machine in one of its U.K. offices. To send a bulletin to all its U.K. offices, Company X would send a single transmission at international cost from its U.S. headquarters to its relay machine in England. The rest of the offices would receive a rebroadcast at local U.K. rates. The numbers of the receiving faxes would rarely, if ever, change, and the cost to send one international fax and four local faxes is much less than the cost of sending five international faxes.

In another case, an agency wants to set up a way to route alert bulletins among its field offices. Each of the field offices periodically generates reports that need to be sent to all the other offices. By placing a relay in one of the offices and programming it to send to every office, a report can be routed by an operator with only one manual transmission.

It is interesting to note that retransmission of a fax by a relay machine does not result in further "fuzziness" or degradation of the transmitted image. This is because the fax is received and stored digitally rather than rescanned for retransmission. Chapter 3 extensively covers the reasons for this and details how you can generate completely "clean" faxes which never touch paper until they are received.

Handsets

Handsets on fax machines are a must-have feature. They allow you to listen carefully to a call, and are very useful when you are trying to figure out why a transmission failed. Often the internal speakers on fax machines are not loud or clear enough for this purpose. The handset can be used as a convenient phone. And for single-call fax-on-demand applications, a handset with touch-tone capability attached to the fax machine is required. On-demand fax is covered later in this chapter.

Simple Manual Operation

All fax machines have the ability to operate manually. Some, however, make this push-button simple, while others make it more like computer programming. Manual operation allows one fax operator to call another and speak over the telephone line before and after a transmission. Manual operation is necessary for using one-call fax-on-demand systems. In the best of cases, manual operation simply involves lifting the handset, dialing a number, talking or interacting with a set of voice prompts, and pressing a clearly distinguished START button. In the worst of cases you have to hit an obscure set of buttons in exactly the right order at exactly the right time to either interact with voice prompts or to engage transmission or reception. Surprisingly, the simpler and cheaper the machine, the easier it seems to operate in this mode. Beware of feature-laden, expensive machines. They are the worst when it comes to manual operation. This is definitely a try-before-you-buy feature.

Paper Cutters

Paper cutters automatically cut rolled paper into pages and are a must for receiving machines. Machines with rolled paper and paper cutters can also be quite useful in proofing fax documents created using the techniques discussed in Chapter 3. A fax that uses a roll and sheet cutter will show you exactly where your page boundaries will appear to a receiver. They will also show you whether a page is short or long when compared to a letter or legal size sheet. A plain paper fax will mask short sheets by printing them on part of a whole sheet. A paper cutter is definitely worth its slight added cost.

Polling

Polling allows you to call a remote fax and pick up a document that is waiting in the feeder or memory of the remote machine. The caller tells the originating (calling) fax machine to "poll" the answering (sending) fax machine, usually by entering a special code on the control panel. The originating (receiving) machine then calls the sending machine and tells it to send whatever document is sitting in the sheet feeder or waiting in memory.

A variation of this technique is called *turnaround polling*. In turnaround polling the originating machine first sends a document then automatically "turns around the line" and receives another document, just as in regular polling. Polling is a feature that is seldom used but which might be convenient for highly specialized on-demand applications.

Security Features

Often security features are added to the typical polling features. Sometimes these require that special codes be provided by the originating operator before the answering machine will engage for transmission or reception. Other security features involve regular transmission or reception. Typically these features block reception or transmission unless specific codes are exchanged between sender and receiver. These are not very useful in applications discussed here.

Error Correction Mode

Error Correction Mode (ECM) is quite useful, but works only with other machines that have it. It works both when sending and receiving. If you have a machine which has this feature—and by some estimates, approximately 70 percent do— then you should turn it on and leave it on all the time. It will help prevent some of the problems with fax transmission, such as scan line dropout, which makes pages of a document look like someone randomly removed thin horizontal strips of paper. Surprisingly, of all the fax modems currently available, only the just announced version of the Gammafax CP, the Gammafax CPi has ECM.

14.4k bps Transmission Speed (CCITT V.17)

This data transmission speed permits transmission at a rate 50 percent faster than most faxes that run at 9,600 bps. Both the sender and receiver must have this feature for it to work. Machines with this feature can be substantially more expensive, and unless you have a closed application and you know that most or all of your machines will have this speed, it may not be useful for transmission with the general fax public. A few fax modems such as the US Robotics Sportster and Intel SatisFAXtion 400 have this speed also. The trend in business use is toward this higher speed. Home users seem to be satisfied with 9,600 bps and lower equipment cost for now.

Computer Interface

An increasing number of fax machines offer this feature, which allows connecting the fax to a computer through a "back channel," typically using a cable and serial ports. The section called "The Multifunction Computer Fax Machine" in Chapter 5 deals extensively with this feature and how it can be used to turn the fax into a very powerful fax-scanner-printer peripheral.

Modified Read (MR)

Modified Read (MR) compression is an improvement over the basic, standard Group 3 fax compression method, Modified Huffman (MH). It provides a reduction in transmission time by greater compression of the transmitted image. MR is a feature available on many higher priced fax machines and a few fax boards. Both the sender and receiver must have this feature in order for it to work. The Intel SatisFAXtion 400 and the Gammafax CP both have the ability to receive these types of transmissions, but only the Intel SatisFAXtion 400 can transmit using MR. The Gammafax CP must be specially configured to allow this.

Modified Modified Read (MMR)

Modified Modified Read (MMR) is an advanced data encoding and compression scheme similar to that used in Group 4 fax. It is a variation of MR compression. It allows faster image transmission through additional compression when compared to both MR and MH encoding. Again, both sender and receiver must have this feature for it to work. Most high-end business fax machines come with this feature, and, 14.4k bps transmission, and ECM. The same comments apply here as applied to 14.4 and ECM.

The Limits of Fax Machines

At some point, you will start to run into the limits of fax machines. Reprogramming speed-dial numbers through a tiny one- or two-line display can be quite time-consuming. Making cover pages manually and attending a machine to be sure the pages feed through correctly can become tedious quickly. Sending a hundred transmissions of a three-page document is simply impractical using an operator and a fax machine. And, beyond a rather limited volume of transmissions, you will encounter the limits of sending over only a single line.

Fortunately, the fax machine, however fancy and feature-packed, is just the beginning of what you can do today with fax technology. As the following section shows, the marriage of computer and fax technology has opened up a whole new way of using fax for your communications needs.

Computer-Driven Fax: The Fax Modem

The integration of computer and fax technologies has ushered in a whole new era in business communications. No other technology provides the kind of easy-to-use, universal, ubiquitous, low-cost, rapid, printed communication that fax does. The popularity of fax machines induced makers of add-on computer products to create the *fax modem,* or *fax board* — a low-cost circuit board that attaches to a personal computer or workstation, allowing it to send and receive fax. These products usually include a set of programs that provide the functions for controlling the board. These programs are what the operator uses to send, receive, and display or print faxes. Some manufacturers have chosen to specialize by providing only control software that can work with the fax boards of many other makers. There are even fax machines that have special computer interfaces that allow them to act both as standalone fax machines or computer-driven fax boards. We will discuss these in greater depth later in the chapter.

Virtually anything you can do with a computer can now also be married to fax. Computers, even small computers, can do just about anything related to automation. They can answer the phone, speak, take a message, listen for and decode pushbuttons, connect with other computers, store or retrieve data in many forms, and, most important, run programs that tell them what to do and how. The computer makes fax completely automatic. For broadcast fax applications, this means completely unattended operation for large numbers of fax transmissions.

To get an idea of what is possible, consider that in a 24-hour period using only a single telephone line and a personal computer and a fax board, it is possible to send over 2,000 separate fax transmissions with just a few minutes of operator effort to start the process. Now consider that there are fax boards that can handle up to 24 lines each. The potential is obvious—that's 48,000 faxes a day with a single personal computer!

Our discussion so far has centered on broadcast fax and how to do it using fax machines. In the next section we're going to look at how to do this using personal computer (PC) fax boards and related equipment. Finally, we will look at on-demand fax, which is not possible with a fax machine.

Understanding How Fax Works

When you work with computers, fax modems, fax broadcast systems, and fax response systems, it can be quite helpful to have a working knowledge of how faxes are sent, received, and stored in a computer. Such an understanding helps in working with all aspects of computer-based fax, including broadcast, on-demand, and production of digital direct fax documents.

Common fax transmission, technically called Group 3 fax, takes place over an ordinary telephone line, the same as used for a normal telephone call. A sending machine calls a receiving machine, the two connect, send and receive a fax, then disconnect. This simplicity of fax masks the details that make such a transmission possible. The following sections show this process in greater detail.

Connecting

When you send a document using fax machines, this is what really happens. The sending machine calls the receiving machine and exchanges messages that help
each side know what to do. During the beginning of the call, known as the *handshake*, the two machines negotiate transmission parameters such as speed, timing, functions to be performed, and other transmission-related information. If you call a fax machine from a regular phone, it answers with a squeal and several chirps, repeating this sequence after several seconds. This is the receiving fax trying to initiate the handshake with what it thinks is a transmitting fax. If you were a fax you would know to respond with some similar sounding noises. You may hear these called DCS (digital control signal), or *training sequences*. By the time the handshake is complete, both machines must agree on how to handle the transmission or they will disconnect without going on to the next steps of the call. Table 3-1 describes in simplified form what two machines do during the course of a call.

Scanning

When the handshake is complete and both machines are ready to transmit, the sending fax begins to scan the paper using a *scan head* inside the machine. This head runs the width of the area where the paper is placed, and contains "eyes" at fixed intervals along its length that "look" at the paper. Each eye determines if the area it sees is light or dark. There are approximately 203 such "eyes" per inch arranged horizontally. Thus for an 8½-inch wide page there would be about 1,727 such "eyes." (If you do the math it actually comes out to 1,725.5 and the difference is due to how fax is specified.) Scanning starts at the leading edge of the paper (that is, the edge that is fed in first) and moves toward the trailing edge.

As the paper moves through the feeder of the fax machine, it is drawn over the fixed scan head. The paper does not move continuously but as a series of discrete, rapid start-and-stop movements. This happens so fast however that it looks continuous. The paper is drawn over the scan head at a rate of approximately 96 or 192 movements per inch, depending on whether the sender is using the *standard* or *fine* transmission mode, respectively. Each time the paper moves then stops, the fixed scan head takes a "picture" of the page, recording whether its eyes see light or dark on that particular horizontal strip across.

For a typical 11-inch-long piece of paper, the sending fax machine takes 1,056 or 2,112 horizontal "pictures" of strips of the page. These horizontal strips are called *scan lines,* and this method of scanning an image using horizontal scan lines is called *rasterizing*. Each page can be thought of as broken up into a table of discrete boxes, or pixels, each of which is either black or white. As the sending fax scans the page, the pixels are transmitted starting with the box numbered 1, until the last box— 1,823,712—is sent. Each box is represented by a single bit of information. Figure 3-2 diagrams this.

SENDER		RECEIVER
Dials receiver	▶	Answers Call
	◀	Start of handshake. Sends answer tone and receiver capability list (max. speed, etc.) and terminal id (TTI or CSID).
Confirms transmission settings for this call (speed, other features) and sends "training burst" to test signal quality.	▶	Responds with "OK" signal if "training burst" is acceptable, otherwise requests a "retrain" at lower speed. End of handshake.
Starts scanning and begins transmission of first page of fax		Receives first page of fax and begins printing.
Handshake again. Signals end of first page	▶	Acknowledges first page, cuts paper or loads new sheet. End of handshake.
(Repeats process for first page until no more pages.)		(Repeats process for first page until no more pages.)
No more pages, signals end of transmission.	▶	Acknowledges end of transmission
Hangs up.		Hangs up.

Table 3-1: Fax Protocol in Simplified Form

1	2	◆◆◆	1,726	1,727
1,728	1,729	◆◆◆	3,453	3,454
◆◆◆				◆◆◆
1,820,259	1,820,260	◆◆◆	1,821,984	1,821,985
1,821,986	1,821,987	◆◆◆	1,823,711	1,823,712

Table 3-2: Scanning Concept

A bit is a binary unit of information. It is represented and transmitted as a one or a zero in a computer, fax, or other digital system. In our example above, the values one or zero are used to represent black or white. Each box can be either black or white, one or zero. Thus, an entire page can be represented as an ordered collection of bits, or ones and zeroes. A byte is merely an ordered collection of eight bits.

Transmitting

The sending machine transmits these scan lines as it moves the paper across its scan head. More sophisticated machines may first store one or more pages in memory before transmitting them. Because the connection between faxes is a serial connection, the data representing each scan line must be transmitted one bit at a time.

A complete set of these scan lines, when placed in sequential order, make up a "photo" of the page. This "photo" is sometimes called a *fax image* or *raster image* and, depending again on resolution (standard or fine), will for a letter size page contain approximately either 1,824,768 or 3,649,536 pixels. You can thus think of a page as being set on a very fine grained piece of graph paper with 1,728 squares across and 1,056 or 2,112 squares down, with each square being either black or white. A fax transmission is basically the sending of each square of the graph paper, one at a time, very quickly, over a phone line.

Receiving

While the sending machine transmits these millions of pixels as ordered sets of scan lines, the receiving machine is busy placing them on paper using a *print head,* which matches the resolution, vertically and horizontally, of the sending machine. That is why on simple fax machines transmission and reception seem to happen in lock step (see Figure 3-3). Memory machines often receive a page into memory before printing it.

Figure 3-3: Conceptual View of Fax Transmission

Compression

The transmission rate between fax machines is typically 9,600 bps, although there are still some machines around which only will work at 4,800 bps. If you do the calculation you will find that sending all those bits should take either three or six minutes a page. We know by experience, however, that a fax transmission typically takes only 45 to 60 seconds. The decrease in time is the product of *data compression*. Built into each fax machine is the ability to compress a fax before sending it and decompress it after receiving and before printing. Using compression before sending and decompression before printing, a typical page can be sent in less than a third the time it would otherwise take. Compression is essentially a machine to machine shorthand that abbreviates long strings of bits that are all the same by substituting some other predetermined code. For example, instead of sending "0000000000" for part of a large white segment of a scan line on a page, the sender might send a single "0" followed by a short code for "repeat this ten times."

Files, Computers and Faxes

The computer comes into play most significantly, of course, when we start using fax modems. When using a fax modem to send or receive a document, the bits or pixels that are normally printed on paper as black or white dots are now stored as data on the disk of a computer, and are called *files* or *fax images*. These files are typically *compressed raster images*, or a stored form of the digital transmission that goes from the sending fax to the receiver. An average typewritten page might take up 30,000 bytes (30k) on a disk when compressed. The same file would take more than 232k as an uncompressed raster file. Using special utilities that are typically provided with fax boards, an operator can view these images on the screen of a computer or print them on a computer printer. What a regular fax would send or receive as a piece of paper, a fax modem sends or receives as a disk file or fax image, since data can be created on disk without printing or scanning paper.

Fax Files and Fax Publishing

If fax modems send and receive fax files yet do not have scanners or printers, how then are images created to be sent or received? In some cases, basic tools for creating memos and other correspondence are provided by the fax modem vendor. These tools, or *utilities,* are programs used for creating documents for sending or displaying or reading documents that have are received. In sending, they create disk files that fax boards then transmit. In receiving, they create files which contain the data or pixels that represent the image of the received fax. These files can then be displayed on a monitor or printed.

But there are other ways to create fax images on disk. Using special computer tools, which are covered in Chapter 5, it is possible to create these fax images directly without ever printing anything on paper to be scanned through a fax machine. You can take output from desktop publishing programs, such as Aldus PageMaker or QuarkXpress, and turn them into *digital direct fax* (DDF) images. These are fax images that are created as files first and then transmitted. The benefit is fax image quality approaching that of laser printers without the usual jagged lines and fuzziness associated with fax. As you read the rest of this book, keep in mind that faxes can exist as disk files first, without ever touching paper until they are printed by a receiving machine.

Computers and Fax Modems

If you think of the fax as a printer, then you can think of a computer with a fax board as a computer that can print to any one of millions of remote printers around the world! It is this ability that makes the marriage of fax and computer such a high-leverage business communications combination.

Exactly What is a Fax Modem?

As previously stated, a fax modem or board is a circuit card that is inserted into or attached to a computer or workstation. It is the "guts" of a fax machine, minus the sheet feeder, scan head, thermal printer, and case and other external hardware. The fax modem contains all the circuits necessary to perform the functions of a fax machine except scanning and printing. It also connects to a telephone line and can dial a number, connect to a remote fax, and transmit. It can wait for a call, and answer and receive a transmission from another fax. To someone at the other end of the phone line, a fax modem is indistinguishable from an ordinary fax. But that's where the similarity ends.

Unless you use a separate scanning device, a fax modem transmits fax images stored as computer files rather than scanning sheets of paper (see Figure 3-4). These stored images are usually documents created using computer-based tools

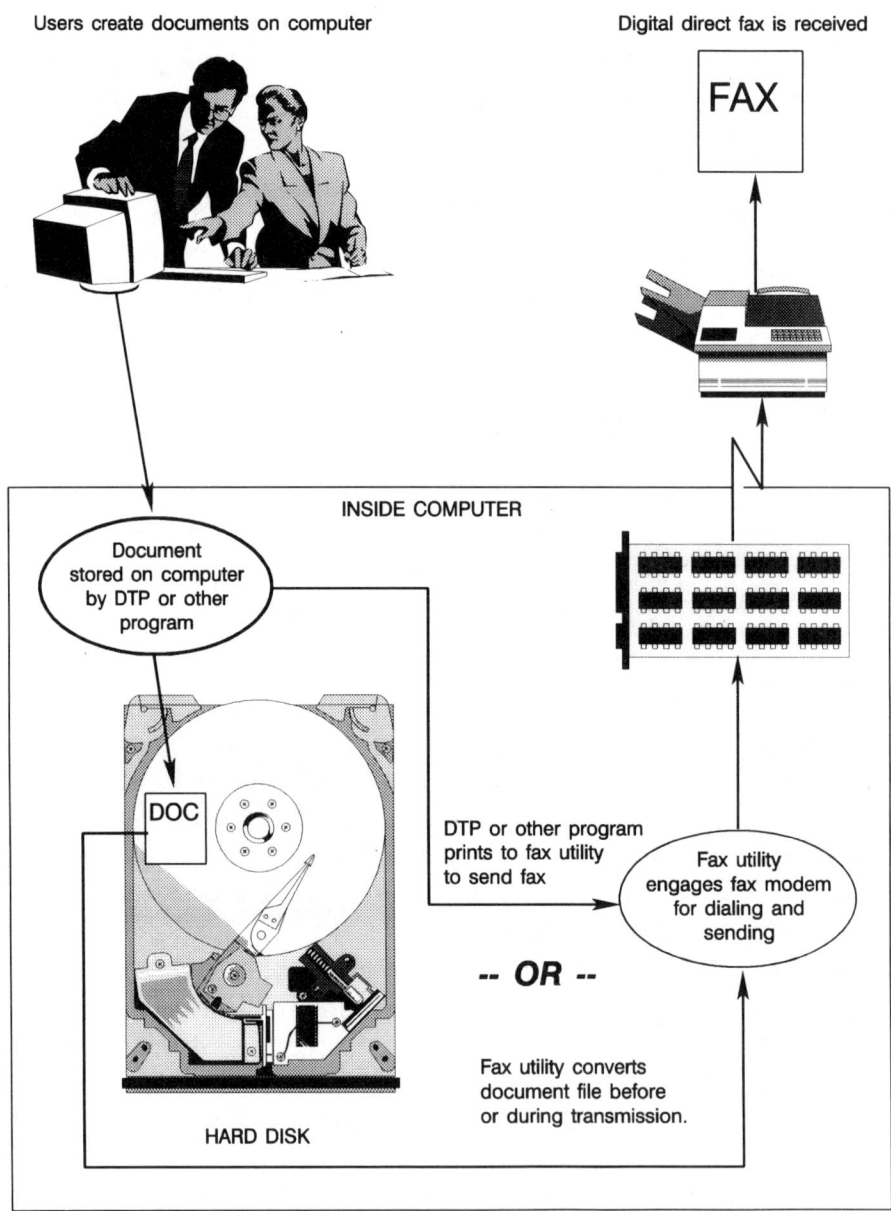

Users create documents on computer

Digital direct fax is received

FAX

INSIDE COMPUTER

Document stored on computer by DTP or other program

DOC

HARD DISK

DTP or other program prints to fax utility to send fax

-- OR --

Fax utility converts document file before or during transmission.

Fax utility engages fax modem for dialing and sending

Figure 3-4: Computer Fax - Sending

such as word processing or desktop publishing programs. When a fax modem receives a document, it is stored in the computer as a fax image disk file. Sometimes these files are called TIF files. (Pronounced "tiff" like a New Yorker might say "Me an duh the wife, weez got into a tiff.") Other popular formats are DCX and PCX. A few products use proprietary formats.

After a fax is received, the computer user (operator) can view the received fax on the computer monitor or print it out on a computer printer. Likewise, the operator tells the fax modem what document to send to what number by using the computer keyboard and one or more application or utility programs (see Figure 3-5).

Even alone, a fax modems can be incredibly convenient and powerful. It can fax documents directly from your computer without having to first print then scan them through a fax machine. The labor and time savings are terrific. Some fax modems can transmit and receive in the background (that is, while your computer does other work) so you can continue to use your computer while sending or receiving faxes. Figure 3-6 shows a typical fax modem.

Perhaps the most powerful feature of a fax modem is the ability to take lists of names and fax numbers in computer file form and use them for broadcasting. Nearly all fax boards come with some kind of utility that provides for storing lists of names and numbers, selecting names and numbers, and sending a document, optionally with a custom-generated cover page for each recipient.

Let's take a look at some fax modems, their features and differences, and how to use them.

"Smart" versus "Dumb" Modems

There are basically two kinds of fax modems: smart and dumb. Smart modems have on-board processors—a simple computer of their own—that does most of the work involved in faxing. These boards offload from the host computer the

Original document is sent by
any fax machine

Users enjoy on-screen viewing

FAX

Fax viewing
utility program displays
received fax on
screen

Fax is received
by a fax modem in
the computer.

Fax board program
puts received fax onto
hard disk

FAX

COMPONENTS
INSIDE
COMPUTER

HARD DISK

Figure 3-5: Computer Fax - Receiving

Figure 3-6: A Typical Fax Modem

work of dialing, connecting, and transmitting or receiving. They rely on the host computer only for higher level functions such as fax and list storage and running background and user-interface programs. Some smart modems, such as the Gammafax CP from Gammalink, come with extensive utility applications including a specialized programming language interpreter for interfacing to the fax modem and its associated programs. Smart modems are better suited for heavy use than dumb modems. Some smart modems, such as the Gammafax CP, can be set up to run multiple lines using multiple fax modems in the same computer.

Dumb modems cost less and can be simpler to install and operate. They are well-suited to the need for occasional desktop fax transmissions and modest broadcasting tasks, but are not adequate for heavy broadcasting. They are limited to a maximum of one line per computer and don't have the added utilities and application programs that make interfacing with other programs easier. Dumb modems rely on the processor of the host computer to handle much more of the work involved in faxing; thus the host processor can be entirely occupied when faxes are coming and going.

One benefit of dumb modems over smart modems is that they usually allow more responsive manual control. This can be important when you need to use a fax modem in conjunction with a handset for operation or maintenance of a broadcast or on-demand system. Dumb modems make it easier to press the on-screen equivalent of the START button. Some smart modems, such as the Gammafax CP mentioned above, do not even have a feature that allows an operator to press START to begin transmission or reception.

Currently dumb modems cost anywhere from $99 on sale at a discount outlet to $170 retail. Smart modems, by contrast, typically retail above $300 and go over $1,000 for a top-of-the-line model. Many of the dumb modems are from off-brand makers, but nearly all seem to be able to accomplish the basic functions of sending and receiving. The real differences you will notice will be in the ease of use, the reliability of communication with other fax machines and modems, and the robustness and convenience of the application software provided. Many fax modems come packaged with fax software such as Bitfax, Winfax 3.0, and FaxWorks which are also available separately. Programs such as these are discussed in greater detail in Chapter 4.

If you plan to set up an operation around a fax modem, you should try out a few before deciding which to buy. By the time you read this, prices are almost certain to have fallen and new products will be available. The information in this book will help you to make a sound choice based on general knowledge.

The vast majority of fax modems are boards that go inside the computer and are made for generic AT-type DOS computers with ISA bus slots (that is, the ubiquitous IBM PC and its clones). There are, however, fax modems for other types of computers. There are now several good fax modem products available for the Apple Macintosh and ,even a fax board for Unix workstations such as those made by Sun Microsystems. The Next workstation comes with a fax modem built in. The Apple Powerbook (a laptop version of the Macintosh) comes with an integral fax and data modem. Fax modems are available for most laptop computers. For laptops that cannot use internal modems, there are external, portable, battery powered models.

Direct Text Conversion

Sending text documents is one of the most common uses for fax, and a situation in which smart fax modems with their own on-board processors can make a big difference. Before a text document can be sent by fax it first must be converted into a fax image. A dumb modem will rely on its application software to do this work. The text-to-fax conversion is done on the host computer, and transmission is a two-step process consisting of conversion followed by transmission. The dumb fax modem application reads the text file and converts it to a fax file on disk. The transmission software then picks up this fax file and transmits it. Depending on the speed of your computer, this can take from a fraction of a minute to several minutes per page (see Figure 3-7).

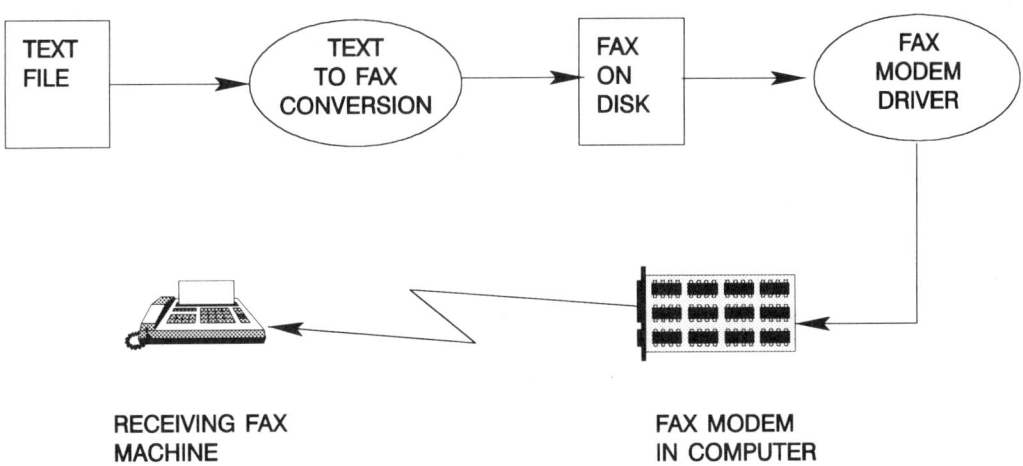

Figure 3-7: Off-Board Text-to-Fax Conversion

Smart modems use their own on-board processors to do this job at the same time transmission is taking place. Thus what was a two-step job is reduced to one step with no waiting. When you tell a smart modem to send a text document, it starts dialing immediately. During transmission it directly reads the contents of the text file and converts it on-the-fly to a fax image using its own on-board processor. It does not need to generate a separate fax file on disk, and does not slow down your computer or require you to wait while it does the conversion before transmitting (see Figure 3-8).

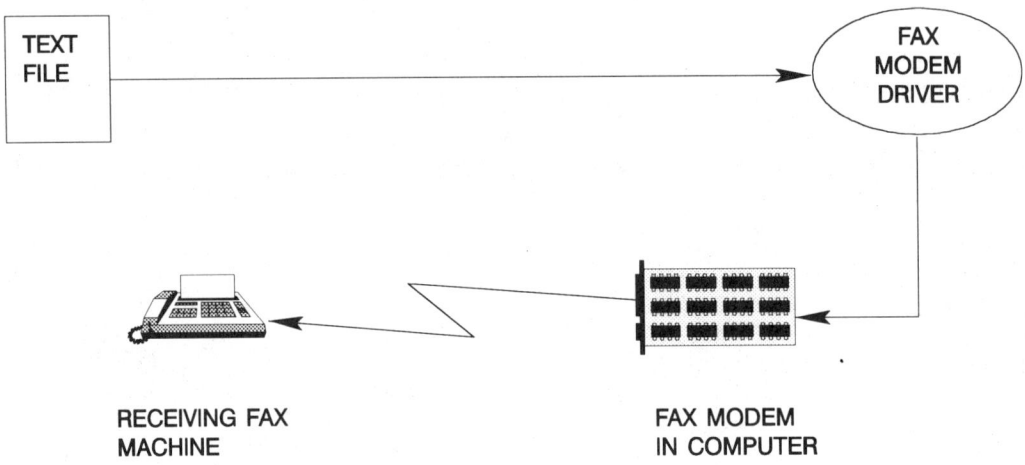

Figure 3-8: On-Board Direct Text-to-Fax Conversion

Foreground and Background Operation

Most dumb fax modems have application software that must run in the foreground, as the only active program on the host computer. To use the fax modem, you must first run the application software, which then allows you to send or receive faxes. No other program can be running while the fax application is running. This can be most inconvenient if you wish to continue using your computer for other tasks while sending or receiving faxes. After all, if the purpose of having an automatic fax broadcast is to save time, how can this be accomplished if you are deprived of using your computer?

Some dumb-modem makers claim their products will operate in a background mode, which is meant to allow continued use of the computer for other applications during fax operation. But this is sometimes unreliable, and greatly slows down the host computer. While operating in this manner may be possible according to the letter of fax makers' instructions, in practice it is not very usable. Most smart modems have acceptable-to-excellent implementations of background operation. This is made possible by the offloading of low-level fax tasks from the host processor onto the fax board.

To use a modem in background operation requires starting an application that stays in memory while other programs run. On DOS computers such an application is called a TSR, which stands for "terminate and stay resident." These applications, which load into your computer's memory and run in the background, operate by using a portion of your computer's processing power when they need to run. They must run whenever the fax modem needs service in sending or receiving a fax. At all other times they run occasionally to monitor for when the fax modem needs service. Generally, the more faxes coming or going, the more these programs will need to work.

Fax Modems and Compatibility

Transmission Compatibility

When Group 3 fax machines first started to sell, problems sometimes occurred when a machine of one type was unable to communicate with a machine of another type. This mostly happened between machines of differing brands.

The truth about the Group 3 recommendation is that conforming to it does not guarantee compatibility. Fax salespersons will claim "any Group 3 machine is compatible with any other," but this simply is not true.

Today, the situation when using a fax machine is much better. No fax is 100 percent compatible with every other machine it might try to work with, but a realistic figure for modern equipment might be 97 percent or higher.

Fax modems share the same compatibility challenges as fax machines. Not all fax modems will work with all other fax modems and fax machines. And *some* fax modems are somewhat more likely to have problems in this area compared to fax machines. When considering purchase of a fax modem, make sure to evaluate its compatibility for your purposes. For example, if you intend to use a dumb modem to upload faxes to a fax-on-demand system, then you should be sure the modem you choose works with that system. If you intend to broadcast faxes to many different kinds of machines (that is, if you are dealing with the public), then you will want to use fax modems that have the most "experience" behind them. This is important because the only way manufacturers have been able to solve the incompatibilities among faxes is to engineer specific features designed to allow their machines to work with as many other brands of machines as possible. And the only way for a manufacturer to get this knowledge has been by finding incompatibilities one at a time and changing the product to fix it.

Here is a practical example of incompatibility. When researching for this book, there was a case in which one fax modem was not able to transmit to a receiving fax machine and another was. In this rare instance, the Intel SatisFAXtion 200 was unable to connect with a remote fax machine. After several failed attempts, the same transmission was tried using a Gammafax CP and was immediately successful. Monitoring the line using the speaker on the Intel SatisFAXtion 200 revealed the Intel fax and the receiving machine were not able to successfully negotiate a handshake to begin transmission. The Gammalink modem had no problem. Again, this a rare situation and not meant to show the Intel SatisFAXtion 200 to be in any way a less than excellent product. Gammalink has gone to great lengths to assure the compatibility of their products beyond a level considered adequate by nearly all other makers. Experience with Gammalink products tends to support this.

It is interesting to note that several months after this test was conducted, it was repeated using new fax software downloaded from the Intel technical support fax-on-demand system. This time the Intel SatisFAXtion 200 tested successfully.

Fax Modem Interface

Fax modems are not like data modems. If you are familiar with data modems then you probably are aware of the nearly universal AT command set. Nearly all data modems seem to use some minimum set of these commands.

Fax modems are completely different. While there is not one standard—such as the AT command set and a simple serial interface—there are three types of interfaces that seem most common and that are supported by most of the fax software on the market.

- *Class 1* fax modems are "dumb." They require the host processor to do nearly all the work, including running the timing-sensitive low-level fax protocols. Class 1 fax modems function best with fax software that runs exclusively in the foreground. When running in the background or when competing for host processor time with other programs such as TSRs they can stifle system performance and cause transmission failures.

- *Class 2* fax modems are the next step up. They offload the majority of timing-sensitive fax protocol chores from the host processor by performing them on the fax board. These modems are "smart." Class 2 modems offer much better host performance and true background operation.

- *CAS* modems support an interface called Communicating Application Specification, which was introduced by Intel Corporation and DCA in 1988. It is an interface that specifies how application programs can work with CAS-compatible fax modems to integrate fax capability with other communications functions. CAS fax modems talk to application programs through a TSR called `casmgr`, which is supplied with the modem. The standard is controlled by Intel and has widespread software support from approximately up to two hundred vendors. It is applicable only to computers using the DOS environment. From a compatibility standpoint, it seems that CAS, like many of Intel's efforts to impose standards, is succeeding. Increasing numbers of applications from fax interface products like Delrina Winfax 3.0 to an increasing number of local area network fax products—seem to be moving toward adopting CAS as their preferred interface.

- *Faxbios* is a new standard developed by the Faxbios Association and was first published in the fall of 1991. It has not been submitted to any standards review groups yet. It supports real-time communications between programs through a procedural API (Application Programming Interface), and is intended to be extended to operating environments beyond DOS. It is unlikely that you would find a retail product advertised as "Faxbios-compatible," but as these fax technologies evolve, Faxbios may become more widespread.

- *T.611* is the first international effort to establish a standard API for fax. It is designed to support Group 3 and 4 fax, as well as other telecommunications services. It was originally developed in Europe, and in 1992 was recommended by the CCITT, an internationally recognized organization that recommends technical standards in the communications field. A draft version of this recommendation will be published in early 1993. T.611 is intended to be platform (computer system)–independent, and is file-based, permitting batch operations. It is a first of a series of CCITT Programmable Communication Interface (PCI) recommendations. Again, it is unlikely that in the next two years such a product will find its way into the market, but it may in the near future.

Transmission Speeds

Nearly all fax modems currently on the market support 9,600 bps transmission, the fastest Group 3 speed available and the speed supported by 99 percent of the fax machine population. Modems that support 9,600 bps also support the slower speeds, such as 7,200, 4,800, and 2,400. But there are a few products that only support lower speeds. There are even a fax modems that supports 9,600 for transmission but 4,800 for reception. The rationale for this is hard to guess. With the products presently available, there is no point in getting a fax modem that does not support 9,600 bps transmission and reception.

Similarly, few fax modems support transmission at 14.4k bps, which is 50 percent higher than 9,600 bps. This additional speed comes at a higher price, and is nearly always coupled with MMR encoding described below. Current market data suggest that over 95 percent of all machines to which you would be likely to transmit a fax do not have 14.4k bps transmission speed capability. As a result, you would be unlikely to benefit from having higher-speed transmission capability unless you knew in advance that a large portion of the recipients of your fax transmissions also had this capability. Furthermore, since this kind of technology seems to get less costly with time, spending less on equipment now and then spending again later may be better than spending more now.

MMR

This is the same feature as described earlier in the section "Using conventional fax machines." Some fax boards, such as the Intel SatisFAXtion 400, have this feature, but others, like the SatisFAXtion 200, do not. The same factors in cost and purchase-timing apply to this feature as to the 14.4k bps transmission speed. This feature will have significant value to you only if you know you will be sending to machines or modems that you can receive an MMR transmission.

Data Modems

Several fax modems also come with data modems as an added feature. The data modem features tend to operate separately from the fax modem, but in some cases they can be configured to support functions such as line sharing between incoming fax and data traffic. The Intel SatisFAXtion modems do an excellent job of managing dual fax and data calls, and also can be configured to support line sharing with voice calls. These dual-purpose fax modems can also be helpful if you are faced with a computer that is limited by available expansion slots.

Fax Application Software and Other Supporting Utilities

Fax modems come with their own special application software and utilities. The Gammafax CP is sold with an extensive set of programs including a specialized command interpreter. Other fax modems come bundled with software that is also sold separately and that can support more than one type or brand of modem. For example, FaxIt for Windows from SofNet, Inc., is bundled with the Intel SatisFAXtion modems, and also supports more than sixty other fax modems.

Whether you use the software that comes with a fax modem or replace it with an add-on package, that software merits close examination. It is central to the operation of the modem and the most important element contributing to fax modem features and utility. Here are some features to consider:

Modem Types Supported

The fax software and modem must be compatible. This means that if the modem has a Class 1, Class 2, CAS, or other interface, the fax software you use must support that interface. Since the fax software is the real determinant of utility and features, it would be wise to consider selecting the software before selecting of the modem.

Sending Options

Look for applications that have delayed transmission, group transmission, and forwarding. *Delayed transmission* lets an operator specify when a transmission should begin. *Group transmission* provides for the selection of lists of recipients for a single document. *Fax forwarding* allows the retransmission of faxes that were received. All these features are frequently used and should be considered important.

On-Screen Viewing

On-screen viewing lets you see a fax image on the screen after you have received it or before you transmit it. This feature can save you time by letting you see a fax without needing to print it on paper. Nearly all fax modem products come with at least one such utility, but some are better than others. Look for products that work quickly (some can take an incredibly long time to display a fax) and have easy-to-use controls for zooming in or out, moving an image around, skipping to the next page in a sequence of pages, rotating faxes that are received "backwards" or "upside down" or interchanging black and white (image reversal).

Editing

Some fax programs allow on-screen editing of faxes. This lets you doodle, write, or mark up a fax before printing or resending it. This can be a real time-saver if your application requires that only a small amount of information be added to received faxes before they are either stored in a fax response system or retransmitted using broadcast.

Text Conversion

Another important feature can be text-to-fax conversion. Nearly all products provide some form of this, but there are wide differences in quality and flexibility. A good product will perform conversion quickly and use an attractive, easy-to-read font. Better products will provide some control over the font used. The best products will perform conversions on-board, as described previously. The Gammafax CP product provides multiple fonts, giving 108, 144, and 216 characters per line and 6 or 8 lines per inch vertical spacing. It can perform both on-board conversion and conversion using a standalone utility program.

Graphics File Compatibility

If you work frequently with graphic files of a particular format, such as PCX, DCX, or TIFF, then you should consider fax software and modems that can directly transmit these files without conversion. Note that the *combination* of software and modem must support the file format you want to transmit. Some software packages permit "attaching" a fax or graphic file to a transmission. This can be a great way to send transmissions that consist of documents from multiple sources. Another desirable feature to consider is the ability to convert received faxes into one of the common graphics formats.

Application Support

If you use a specific application or operating system feature, such as DDE links, you will want to make sure this feature is supported. Your choice of modem and software can be guided by the other programs you already have.

Programming Language

This is a highly desirable feature for those looking to handle large numbers of transmissions and integrate their fax broadcasting with databases for list management and other purposes. A good example of a simple yet powerful programming language for a fax modem is the Gammafax GCL utility program, provided with Gammafax CP modems. The GCL utility is a simple program interpreter and will process commands stored in a text file. Commands in GCL

COMMAND	DESCRIPTION
ANSWER n	Wait for incoming call
AUTOPOLL n	Disable/enable turnaround polling
CALLS n	Set number of re-dial attempts
CDTIME n	Wait n seconds for carrier
$COUNTER n [m|op]	Reset dynamic counter n
COVER n filename	Set cover sheet transmit on/off, cover filename
CSID number	Set subscriber Identification (CSID)
DATE	Report system date
DEBUG n	Set debug print level control to n
DELAY n	Wait n minutes between retries
DELETE list record	Delete transaction record from list
DIAL =n number	Dial number with n tries
DOEXIT	End DOWHILE loop
DOS command	Execute DOS application
DOWHILE ... ENDDO	Continue conditional command processing
ENDTIME h:m:s...	Set an absolute finish time for a transaction
EXECUTE filename	Execute a DOS program
EXIT n	Exit, return to DOS, set error level to n
GET list record	Get a transaction from a list
HEADER n text	Set header transmit on/off, data for header
HELP	Give program help
HOLD h:m:s...	(see STARTTIME)
IF ... ELSE ... ENDIF	Conditional command execution
INQERROR	Report the error code of the transaction
LIST filename	Send a list of files
LOG n filename	Log messages at level n to filename
MODEMID name	Select specific modem for transaction

Table 3-3: GCL Fax Programming Language

COMMAND	DESCRIPTION
NOTIFY n	Set notification level for transaction
NSF n field	Set nonstandard facilities field on or off
NUMBERS filename	Dial using the numbers in a file
PRIORITY n	Set priority level for transaction
PUT	Put a transaction on a list
QUEUE filename	Specify default queue filename
QUIT n	(see EXIT)
RATE bps	Set maximum bit rate for transmission
RECEIVE filename	Specify file name for received images
RECOVER n	Select a retry procedure
REMARK comment	Write comment to log file
REPORT oper fmt	The format for record printing to screen and logs
RESET	Reset to default values
RETRY n	Set number of times to attempt transaction
RUN fname	Execute commands from a file
SECURE n	Check answering CSID against phone number
SEND filename [options]	Specify file(s) for transmission
SETERROR n	Set error level for transaction
STARTTIME h:m:s...	Schedule a transaction for later time/date
STATUS command	Report current value(s) of command
SUMMARY filename	Report transaction in memory to a file
TIME	Report current system time
USERID name	Set user name for transaction
WAIT n	Suppress execution for n seconds
XDELETE n	Allow filename overwrite in XFER mode
XFER n	File transfer toggle
XLIST n filename	Send a list of files

Table 3-3 (Continued): GCL Fax Programming Language

provide control over practically every aspect of fax transmission. Table 3-3 summarizes the GCL commands, and should give you an idea of what the program can do.

Here is an example of how GCL can be used. The following list of commands sends the file faxart.tif to three separate recipients by telling the fax modem to dial the three numbers next to the dial commands. In each transmission the same fax is sent. The program also tells the modems to start sending at 11pm and if a fax cannot be delivered due to a problem such as a busy line, then retry up to three times at ten minute intervals.

```
starttime 23:00
send faxart.tif
retry 3
delay 10
dial 1-415-776-0615
dial 1-415-555-1212
dial 987-6543
```

These commands might be stored in a file named testgcl.txt. To start these transmissions, an operator would run the GCL program as follows:

```
gcl testgcl.txt
```

In some examples at the end of this chapter, we will get into more detail on how you can use GCL programming to implement a database driven broadcast fax system and a fully automatic fax relay.

Printer Emulation

A fax modem with this feature emulates a printer by taking output normally intended for a printer and converting it to send as a fax instead. This is very convenient, and opens up an easy path for generating digital direct faxes without any explicit conversion step. (More on this in Chapter 4.) The *Intel SatisFAXtion* products support HP LaserJet (PCL5) and Epson printer emulation. This means that if you have an application that can print to either of these common printers, the SatisFAXtion product will automatically convert that output into a digital direct fax and transmit it.

Scanner Support

Support for a desktop scanner like the Hewlett-Packard ScanJet can be a major convenience if you must scan in paper documents as part of a broadcast. The majority of fax software does not support scanner input. A way around this is to use scanner software to scan the document into a file format such as PCX, DCX, or TIFF, that is compatible with the fax modem you intend to use. Some models of the Intel SatisFAXtion modems have a scanner port that will work with certain hand scanners, but these are only useful for capturing parts of pages. If your application is DTP-based and likely to be scanner intensive, you should consider carefully the equipment described in the section on "The Computer-Fax Machine Connection."

Operability

Not all fax modem and software combinations will function smoothly or be able to handle the workload you may wish to impose on them. How smoothly and reliably you can fax on a personal computer depends both on the type of modem used (Class1, 2, CAS, or other) and on the ability of the fax application program to work with it. Not all application programs will provide the same quality of operation given different operating systems (Windows, DOS, and so forth) and modem types. Performance can differ significantly, for example, between operation under DOS versus under Windows.

An example of differences in operability can be seen between how Winfax operates with a CAS modem and a Class 2 modem. Operation is very smooth with the CAS modem because the `casmgr` program that resides outside Windows handles the low-level protocols and their time-sensitive demands, relieving Winfax of this task. With a Class 2 modem, Winfax must routinely take over the host computer to assure it is available to meet the timing needs of the low-level fax protocol. While transmission or reception is taking place, a Windows user will get the hour glass icon, which indicates the computer is busy, much more often with a Class 2 modem.

This improved operability with a CAS modem comes at a price. The CAS modem requires a driver and a TSR to load prior to starting Windows. This reduces the amount of conventional memory available to other DOS and Windows programs. For the Intel SatisFAXtion modems, the driver takes 3.9k. The TSR `casmgr.exe` takes 5.4k of conventional memory if configured to use EMS memory and 65.9k if restricted to using conventional memory. Loading a TSR to run a fax modem trades memory for improved performance .

Depending on the size of the TSR, this tradeoff may not be acceptable. For example, the fax dispatcher program, `gfdcp.exe`, which drives the Gammafax CP fax modems, uses 90k of memory when loaded to run as a background TSR. This can be far too much if your conventional memory is already loaded with drivers and other TSRs, and you have applications that require a lot of conventional memory in order to run.

If you are using Windows 3.1, one way to work around this is to run `gfdcp` as the foreground task in a background-enabled DOS window. This has several benefits over the prescribed method of loading `gfdcp` as a TSR under DOS before starting Windows. It saves about 90k of conventional memory; it is possible to start and stop the program easily without rebooting; and diagnostic messages showing call progress and modem operation can be easily monitored or ignored.

Gammalink will tell you this is not supported and may even claim it will not work. However, actual use shows that it works well if you limit the number of boards to one or two, use a machine at least as powerful as a 386/33, and have a fast disk (15 ms or better), about 2Mb of disk cache, and 8Mb of real memory. It's possible to get by with less, but if you intend to do reasonable amounts of work while sending and receiving faxes on one or two boards, using less than this configuration is probably a bad idea.

If you want to try this, you will need to set two sets of parameters in Windows 3.1. One is for the 386 Enhanced mode settings, and the other is for an application-specific PIF file. Also, it is important to note that no applications should be allowed to run exclusively, or this setup will not work. Settings not specified are the same as the defaults.

The settings for 386 Enhanced mode are shown in Table 3-4.

PARAMETER	VALUE
Windows in Foreground	128
Windows in Background	64
Minimum Time slice	16

Table 3-4: 386 Enhanced Settings for GFDCP in a DOS Window

The PIF settings are shown in Table 3-5.

PARAMETER	VALUE
Video Memory	Text
Memory Requirements	128k, (0 EMS, 0 XMS)
Display	Windowed
Execution	Background
Background Priority	50
Foreground Priority	50
Lock Application Memory	Yes

Table 3-5: PIF Settings for GFDCP in a DOS Window

Multiline Fax Modems

Multiline fax modems can send multiple fax transmissions simultaneously over multiple phone lines from a single machine. They do this by supporting multiple lines per fax modem (that is, more than one fax per board), multiple boards per machine, or both. The operation of multiple boards per machine must be supported by the software used. Most fax software packages are intended to operate only a single line. If multiple lines are what you need, be sure to pick both software and fax modems that work together and will operate multiple lines simultaneously. For example, the Gammafax CP fax modems will operate up to sixteen lines per machine and comes with software that will support multiline operation. By contrast, the Intel SatisFAXtion 200 and 400 modems will support multiline operation, but most software packages designed for it will not.

Transmitting over multiple lines is important when you have a lot of transmissions to send out within a given time limit. Let's take an example. Suppose you had to send an overnight fax to seven hundred people, and each transmission was expected to take two and a half minutes. Also you want to send between the hours of 11 pm and 7 am only, to take advantage of lower transmission costs. Using a single fax modem, it would take more than twenty nine hours to send all the faxes. You only have eight hours. With a four line fax broadcasting system, you could complete all transmissions in the required time.

Specialized Fax Broadcast Systems

Specialized equipment is available for fax broadcasting. This equipment is specifically set up to handle the job of fax transmission and can save you the time and experimentation involved in installing and adjusting the setup of a computer and one or more fax boards. Also, depending on the nature of your work, it may be advantageous to have a special machine dedicated to the job of transmitting fax.

Two systems that are based on the Gammafax CP modems are the FAXBLASTER from DBC Associates, and the TurboFax from the Turbo Group. Both are essentially DOS-based computers preconfigured with fax modems, software, and programming to make them ready to run. These systems typically handle from one to sixteen phone lines, and are suitable for most business applications. There are other makers of highly specialized large-volume broadcast systems, but these are generally custom built for use by service bureaus. FaxBACK, Inc., a maker of fax-on-demand systems has a feature that allows its to be used as a part-time fax broadcasting system.

There are also specialized software packages and systems intended for connection to LAN-based systems. These typically are intended to allow sharing of fax modems by computer users attached to a LAN, but often may also be used for broadcasting. Many of these products will support multiple fax modems per LAN fax server. Davidson Consulting publishes a report describing in detail nearly every product available in this category (see the Appendix for contact information). If your use of computers is already LAN-based and you are considering integrating your fax broadcasting operation into your LAN, then you should look into network fax server technology.

Scanners

A scanner is a peripheral device that attaches to a computer and functions much like the scanner part of a fax machine. Scanners take pages of documents and capture them as raster images (files) on the disk of the computer. Used in conjunction with a fax broadcasting system, a scanner can be a much faster way to handle printed documents.

Most fax broadcasting systems, whether you use fax boards in your own computer or purchase a system prepackaged, do not come with any way to feed in paper documents for retransmission. If printed documents are to be transmitted, then they must first be fed into the fax broadcasting system, typically by transmitting them in via a fax machine. Thus, a regular fax machine is used as a scanner. This can be a problem if the broadcasting system needs to take input while it is also transmitting. To assure availability, either an entire line must be dedicated for input, or input must wait until the broadcast machine is "between jobs" and is not transmitting.

A separate, attached scanner makes this task much easier since it can accept input for the fax broadcasting system without needing to transmit the document using a separate fax, dedicating a line to inbound traffic or waiting for outbound transmissions to clear.

The Hewlett-Packard ScanJet product line is well suited to the task of scanning for a fax broadcast system. It is directly supported by the Gammafax CP utility programs, and is also supported by a wide range of desktop publishing software programs. It can serve double duty as both fax broadcasting system input scanner and an image capturing device for desktop publishing (see Figure 3-9).

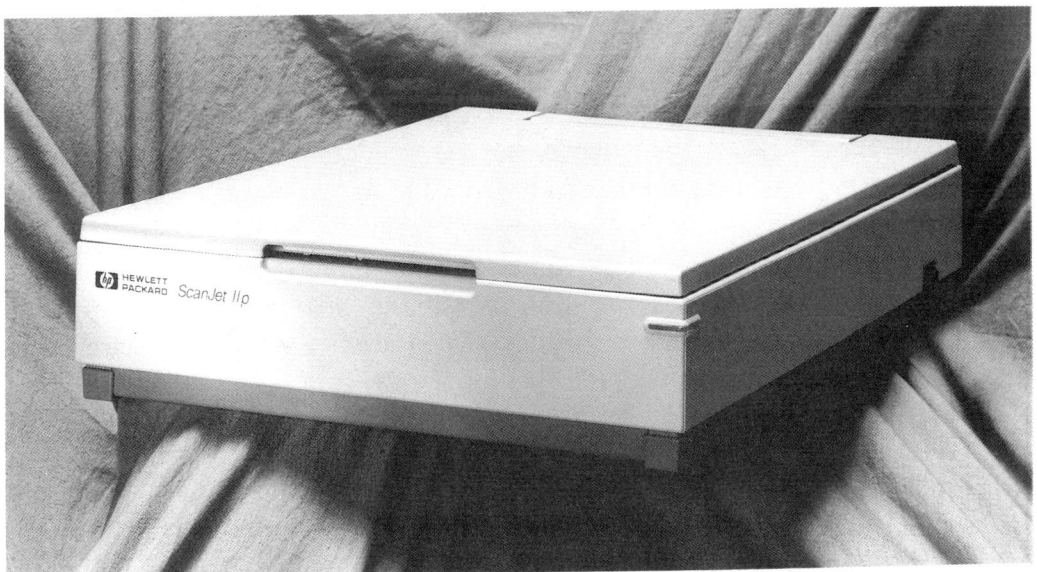

Figure 3-9: Hewlett-Packard ScanJet Scanner

Let's look at how a scanner might enhance a fax application in business. A large law office is constantly transmitting faxes to clients and other offices.

Often, documents must be faxed to four or more recipients. Because of the volume of transmissions, the firm employs a bank of several fax machines and operators to feed them. The operators spend time making cover pages, assuring that the documents feed correctly, and making sure the documents transmitted. If the receiving numbers are busy, the operators need to supervise the retransmissions.

This process is greatly streamlined with a fax broadcasting system with an attached scanner. Documents to be transmitted are fed into the broadcasting system via the scanner and stored in the fax broadcasting system. (If you choose such a system, be sure to purchase the document feeder attachment. It is a real time-saver for multipage documents.) This is faster than transmission via fax because the scanner is connected directly to the broadcast system. The operator can verify that the pages were scanned correctly by viewing them on the fax broadcast system screen prior to transmission. There will be no need after that to make sure the paper feeds through the fax machine correctly for each separate transmission.

A ten page document scans into the system in about a minute; thus the documents are returned immediately to the sender while they wait. The broadcast system has been configured to store the names and fax numbers of all expected fax recipients—clients, firms, title companies, government offices, and others. The operator picks these names and fax numbers from a list using a menu interface, and presses a button to tell the broadcast system to begin sending to the chosen list of recipients. When a name and number are not in the stored list, the operator just types them in.

The broadcast system automatically begins transmission and reports the status of each transmission after it is complete. Cover pages are generated automatically and include the letterhead and contact information for the firm. If all lines are busy with other faxes, the broadcast system queues the new faxes and waits for a line to become free. If the system encounters a busy fax number, it places a temporary hold on that specific transmission and moves on to the next one in line. After a period of time it automatically retries any busy numbers. This all takes place without operator intervention.

Service Bureaus

You can do fax broadcasting with a minimum of equipment if you employ a service bureau to do the transmissions for you. There are many service bureaus that will transmit your faxes using their equipment for fees ranging from twenty cents to a dollar per page. A service bureau may be as small as the copy show down the street or as large as the international office that can send hundreds, or even thousands of faxes in minutes. Many of these broadcast services are provided by the same companies that sell long distance telecommunications services, and are sold under easily recognized brand names: SprintFax, MCI Fax, and AT&T Enhanced Fax. Other large players in the business include companies like Cable and Wireless and Swift Global Communications.

Here is how a typical fax broadcasting service, Swift Global Communications, works. You, the sending customer, prepare a document to send, and transmit it by fax to a toll-free number using either a fax machine or fax modem. Your list or recipients is established in advance and transmitted to Swift separately. Upon receiving your faxed document in their computers Swift begins broadcasting your document to the people on your list. Each transmission is individually addressed using a cover page. At 0500 New York time, Swift sends to you a report detailing the status of each transmission. Swift's fax transmission system will attempt to reach busy fax numbers for up to twenty four hours. For customers who wish to transfer ASCII text files directly into the broadcast service for transmission, Swift provides its own menu-driven communication software to you. You get a bill once a month. The cost is approximately twenty cents off-peak and forty cents on-peak per minute.

Why use a service bureau? There are several advantages, even if you intend to use your own equipment for broadcasting.

- *Lower investment.* You don't have to buy or operate any equipment. This saves your capital for other important jobs, such as marketing or product development. It also saves you time and effort because there is less to learn and handle. Your training cost is reduced. The tradeoff in using a service bureau is that the cost per transmission is likely to exceed the cost of transmitting with your own system.

- *Higher capacity.* Service bureaus typically have very high capacity because of the volume of faxes they send. If you must send large numbers of faxes within short periods, a service bureau may be far more economical than purchasing such high-capacity equipment for yourself. If your needs are for high volumes of transmission over very short periods of time—say 10,000 faxes in an hour—a service bureau may be the only way to economically get the transmission capacity you need. Even if you have your own equipment, using a service bureau may still be effective for those occasions where the workload exceeds its capacity.

- *Backup*. If you have a critical application involving fax broadcast, a service bureau can function nicely as a backup system. It can be used in the event of equipment failure or operational error. Io Publishing, which publishes the daily fax newspaper *BioWorld Today* uses a FAXBLASTER from DBC Associates for most routine transmissions. However, the company has maintained AT&T Enhanced Fax service as a backup by periodically sending part of its daily transmissions or an occasional promotional transmission over this service.

- *Lower international costs.* Another good time to consider service bureaus is when you frequently send faxes internationally. Some service bureaus operate their own fax transmission networks, which can actually provide you with lower per transmission costs than if you were to send the faxes directly yourself. If much of your fax traffic is international or you anticipate that it will be, it's best to shop around early and get an idea of service bureau pricing before deciding on equipment. One word of advice: while most firms are competitive on domestic rates, international rates can vary widely and depend on many factors such as destination country. Assume nothing and shop carefully.

Let's look at an example involving lower investment and higher capacity. Say you have six hundred faxes to send out and they must all be received within an hour. Let's also say that each fax takes exactly one minute to transmit. If you were to set up your own fax transmission system each line could send sixty faxes an hour. (This is a simplified example.) If your equipment had ten lines, you would have just enough capacity to complete all transmissions in one hour. If each line costs $1,000, then you would have a $10,000 investment to make. A service bureau, however, would eliminate the need to buy this equipment. A large service bureau should have more than enough capacity to send your transmissions within an hour.

Selection Factors

You might think that pricing and features in the fax transmission business would be simple, but because it is a new business, there are a few things you will want to examine closely when comparing services and planning your operations. Here is a brief checklist of things to watch in the process:

Billing Unit

The unit by which services are charged is far from obvious. A buyer should thoroughly understand the parameters involved in transmission charges. What is the unit of billing? Some companies bill by the minute and others bill by the "page." But what is a page? Do they mean a letter size page? Legal size? International A4? One large and well known company when pressed for a definition said a page is a separate sheet of paper that does not exceed "40k." When asked what was meant by "40k," they replied that it was some internal measure their machines made. When asked how one could determine if a page was "40k" or not, they said that only they could tell! In their defense, the inquiry was made when their service was fairly new. Let's hope they have a better answer by the time you read this.

Billing Increment

Most minute-based billing is in increments of single minutes. This means that if you have a 61-second transmission, you will pay for two minutes of service. Some services offer billing in tenth of a minute increments. Thus, 61 seconds would be charged as 1.1 minutes. The more transmissions you make, the more this adds up to a big difference in cost. As for "page-based" billing, a "short" page, such as half a letter-size sheet, would still count as a full page even though it was much smaller and less costly for the service provider to transmit.

If your transmissions are well characterized, then it is possible that a billing scheme that appears more expensive at first actually works out to be lower-cost in the end. Take the following example. Your transmissions always consist of a two-page document that takes ninety seconds to transmit. Service Bureau A charges thirteen cents a minute with a one-minute billing increment. Service Bureau B charges fifteen cents a minute with a six second (0.1 minute) billing increment. The cost per transmission of using service A would be twenty-six cents, the charge for two minutes. The cost for service B, which appears at first to be more costly, would be twenty-two and a half cents per transmission. That's a cost savings of about 13.4 percent.

Transmission Reports

These reports are produced by the service bureau after a transmission job has been completed. They are important because they tell you what faxes were delivered and when. More important, such a report should show you which faxes were *not* delivered and why. Quality and detail in transmission reports varies significantly between providers, so be sure to ask for a sample copy.

Delivery Time Guarantees

If you have an application that is time-critical, be sure to discuss this with a prospective service bureau and obtain in writing their commitment that they will in fact be able to deliver all your faxes within your specified time limits. Don't take it for granted that all service bureaus have enough capacity to meet your needs, or that they will operate in such a way as to provide a certain level of service unless it is specified in a contract.

List and Document Handling

A service bureau should provide an efficient way to transfer documents and lists. Some service bureaus handle the process with customer service staff. Others provide automated or semi-automated means of using computers, faxes, or voice response for handling these tasks. Ask to see all of what is available and pick the service that best matches your needs.

Customer Service and Support

One of the most important reasons to use a service bureau is to save your own time and effort. For that reason, fast, competent service is essential. If you have any doubts, ask for references. Try and find out what the vendor does when things go wrong. This is much more important than knowing that everything is fine when there are no problems. If you are sending a lot of faxes, then at one point or another you will experience problems and require assistance. Examples of such problems include a sudden increase in transmission failures or a rapid drop in the quality of received faxes.

Example Broadcast Applications

The following broadcast applications are presented to show how broadcast fax can be used in actual business situations.

Database Driven Fax Broadcast

Fax broadcasting can be an efficient way to handle business situations where you need to contact more than a few parties in a short period of time. This example shows how broadcast fax is used to support an important part of the buying process: getting requests for information or quotation out to vendors. In the following example a database of vendor information is used to drive fax broadcasts. The database is used to determine the list of appropriate vendors. Each vendor receives a form letter which is a request for information or quotation and has the option to respond by fax or by telephone.

The Database

The database in this application is Paradox 3.5 from Borland International running on a generic 386 AT-type computer. Records are kept that contain vendor information, including the types of products sold, a contact name, and the fax number.

Fax Board

The fax board in use is a single Gammafax CP. The program that supports the fax boards runs in a background DOS window under Windows 3.1. The user interface or menu program runs in a separate DOS window.

How It Works

A form letter is written. It asks for quotations or information regarding products or services of a specific description. This letter is composed using a text editor and includes special control codes that tell the Gammafax CP to attach graphics such as a letterhead and a signature. The letterhead and signature were created using desktop publishing (DTP) tools. The text of a sample letter is shown in Figure 3-10, and the fax produced from this text is shown in Figure 3-11.

```
@/c:\gfax\images\lhead.tif@/
*** ATTENTION FAX OPERATOR PLEASE DELIVER IMMEDIATELY ***

From: Phil Sih
To:   Sales Department
Re:   Request for Quotation

WANTED -- 286 Motherboards
We are an original equipment manufacturer looking to purchase current
in-production low-cost 286 motherboards for use as components in our
products.  The boards must meet the following specifications:

REQURED:
286 processor
Space for 2M on board DRAM
Boots *without* monitor card or keyboard attached
Standard form factor (size, placement of holes, etc.)

OPTIONAL desirable features or acceptable limitations:
AMI BIOS
On-board IDE FD and HD controller
May be any clock speed.
Only 3 expansion slots required.
(May have only 2 slots if on-board IDE FD and HD controller present.)
Manuals and other "end user" materials only needed for evaluation.

PRICING:
Please provide complete volume pricing.  Price will be a determining factor
(but not the only factor) in determining the winners.

OTHER FACTORS:
Other factors which will influence our purchase decision include: Product
availability (current production a plus), Terms and conditions of sale.

PRODUCT IDENTIFICATION:
Be sure to identify your product by your part or stock number.  Most
motherboards do not come with any identifying marks.  We need to be able to
easily identify your product if we are going to purchase it.

Sincerely,
@/c:\gfax\images\philsig.tif@/
Phil Sih
Managing Director
DBC Associates
```

Figure 3-10: Text of Sample Letter for Database Fax Broadcast

---

DBC	**DBC Associates**	1 Daniel Burnham Court	TEL:	415-776-6227
		Suite 809	FAX:	415-776-0615
		San Francisco, California 94109		

---

xxx ATTENTION FAX OPERATOR PLEASE DELIVER IMMEDIATELY xxx

From: Phil Sih
To: Sales Department
Re: Request for Quotation

WANTED -- 286 Motherboards
We are an original equipment manufacturer looking to purchase current
in-production low-cost 286 motherboards for use as components in our
products. The boards must meet the following specifications:

REQURED:
286 processor
Space for 2M on board DRAM
Boots xwithoutx monitor card or keyboard attached
Standard form factor (size, placement of holes, etc.)

OPTIONAL desirable features or acceptable limitations:
AMI BIOS
On-board IDE FD and HD controller
May be any clock speed.
Only 3 expansion slots required.
(May have only 2 slots if on-board IDE FD and HD controller present.)
Manuals and other "end user" materials only needed for evaluation.

PRICING
Please provide complete volume pricing. Price will be a determining factor
(but not the only factor) in determining the winners.

OTHER FACTORS
Other factors which will influence our purchase decision include: Product
availability (current production a plus), Terms and conditions of sale.

PRODUCT IDENTIFICATION
Be sure to identify your product by your part or stock number. Most
motherboards do not come with any identifying marks. We need to be able to
easily identify your product if we are going to purchase it.

Sincerely,

Phil

Phil Sih
Managing Director
DBC Associates

Figure 3-11: Fax of Sample Letter for Fax Broadcast

```
-------Name---------------Wfax-----
3Q Computer Inc.     |  408-248-8987
AMT International     |  408-944-9801
ASA Computers        |  408-988-0359
ATR Systems          |  415-683-8979
Anova                |  408-943-9660
Aroma Computer Syst  |  408-736-4356
Asia Source          |  510-226-8858
Belmont Systems Inc  |  415-598-0186
Bozeman Marketing G  |  408-263-6763
C&J Computer         |  408-727-1209
Central Computer Sy  |  408-241-0390
East Gate Micro Inc  |  408-262-3340
Elitegroup Computer  |  510-226-7350
GT Computers Inc.    |  510-659-1309
Gems Computers Inc.  |  408-473-0826
Grey Microsystems    |  408-436-1108
Hokkins Systemation  |  408-436-3021
King Star Computer   |  510-659-8644
Legend Microsystems  |  415-969-8931
Libre Service Compu  |  510-438-0489
Libre Service Compu  |  408-727-1760
MCC Systems Inc.     |  408-255-4348
MW Computer          |  510-659-8526
Microland Electroni  |  408-441-1767
Microline Computers  |  510-770-1912
Mitsuba              |  408-441-8558
Nexus Systems        |  516-338-6626
Orx International    |  510-490-8683
Roger Willis Comput  |  415-573-1566
Schwab Computer      |  408-245-3103
Sunlight Computer &  |  415-692-0712
Super Link, The      |  408-745-1609
Systender Internati  |  408-732-7030
Texas Microsystems   |  713-933-1029
Tower Computer Inc.  |  408-288-7120
Tower Computer Inc.  |  408-929-8638
```

Table 3-6: Table of Numbers and Recipients for Fax Broadcast

The text that makes up this form letter tells the fax board first to get the graphic contained in the file `c:\gfax\images\lhead.tif`, and transmit it as the first part of the fax. The letter then says to take the text after the first line, convert it to a fax, and send it next, without making a page break. Then, toward the bottom of the letter, fax the signature by, appending the graphic stored in the file named `c:\gfax\images\philsig.tif`, and finish by converting the last few lines of text into a fax.

A query is run on the database to determine all appropriate vendors for this request. In this case, we're interested in getting all vendors that sell *motherboards*, a board that contains the central electronics of a specific kind of computer. The query gives us the list of vendors as shown in Table 3-6.

This is a list of thirty five vendors. To have the fax modem send our form letter to this list of recipients, we will need to use GCL, the programming language feature of the Gammafax CP product. To use GCL, we will need a text file containing the program to be processed by GCL . A convenient way to create this program is to have the Paradox 3.5 database place the list of vendors on disk in a file format that we can then process using GCL. This is simple, since all we need is a list of numbers placed in a file named something like `numbers.txt` as shown in Figure 3-12.

```
408-248-8987
408-944-9801
408-988-0359
415-683-8979
408-943-9660
408-736-4356
etc...
```

Figure 3-12: Sample Fax Broadcast Number List File

We then create a GCL program that looks like the one shown in Figure 3-13.

```
send letter.txt
start 23:00
numbers numbers.txt
```

Figure 3-13: Sample GCL Broadcast Program

This short program tells the fax modem to send the form letter to the list of numbers contained in the file numbers.txt starting at 11 pm on the present day. Thus the 35 faxes would go out during the lowest rate periods and arrive on the desks of the vendor salespeople the next morning. This method of communicating is very effective. People who are motivated to sell their products will call or fax back in response the next day. In about ten minutes you can communicate with thirty five vendors and accomplish what would normally take a full day or more of work on the telephone.

Solving the Addressing Problem

In Chapter 2 we examined the publication *BioWorld Today* from Io Publishing. It is interesting to examine that application in detail because it illustrates one way to very conveniently and economically solve the addressing problem.

Since many fax machines are located in a mail room or similar central facility, fax transmissions often require a cover page to tell the receiving operator who gets the fax. *BioWorld Today* must have the name of each subscriber and some delivery instructions on each of its transmissions.

The method the company uses eliminates the cover page entirely and clearly marks each page in the transmission—typically two to four pages—leaving all this work to the computers! Here is the solution.

Unlike the above example, where we had the database kick out a file that was just a list of names, the *BioWorld Today* database (Filemaker running on an Apple Macintosh) kicks out a complete GCL program that is then processed using the GCL utility. The result is a line of text at the top of each page that appears as shown in Figure 3-14.

```
02/06/93 23:04:22    PLEASE DELIVER TO:->              TO:PHIL SIH      Page   1
```

Figure 3-14: Close-up of Header from BioWorld Today

For *BioWorld Today* this is better than a cover page. It costs about one tenth as much to transmit resulting in an approximate savings of about five cents per transmission. The header appears on every page, so it is less likely that pages will be lost. And it clearly personalizes each copy of the newspaper. What a great feature for customer relations!

Here is how it is done. In a file called `header.gcl` there are GCL commands as shown in Figure 3-15. These commands tell the broadcast system to send the file `us001.tif` starting at 11 pm.

In a file called `doit.gcl` the information for each subscriber is included as shown in Figure 3-16. Note each subscriber has a "header" and a "dial" line.

The Filemaker database was programmed to generate this file as a text report on disk.

```
recover 2
csid PLEASE-DELIVER-TO
starttime 23:00
send us001.tif
```

Figure 3-15: BioWorld Today GCL

```
run header.gcl
header 1 Joe D. Subscriber
dial 1-415-776-0615
(and so forth for the entire subscriber list)
```

Figure 3-16: BioWorld Today Batch Program

When *BioWorld Today* is ready to go to press, it is stored in a series of files named us001.tif, us002.tif, and so forth. An operator takes those files and copies them onto the FAXBLASTER fax broadcast machine, along with the above GCL files. Then with a single command line: gcl doit.gcl, the whole batch is then set up to run at 11pm.

Using the CSID and HEADER fields for the routing information causes the GCL utility program to place these fields in an internal data structure called the *fax queue* which lives on disk in a file called gfax.$qu. During the course of actual transmission, the fax manager program, gfdcp.exe, which runs in the background and services the fax modems, passes this information to the modems. The modems then create the header as it appears in Figure 3-14, attaching it to the top of each page transmitted. Although in this case the default header style was used, you can customize the header by telling the modems to use different fields that can be passed via the GCL utility. These fields include a few dozen standard fields and several user defined fields. This combination makes possible practically any kind of header you can imagine. The Gammafax Reference Manual that comes with the Gammafax CP modems gives complete detail on programming this feature and a whole host of others.

A High-Speed Fax Relay

A local police department generates approximately a hundred incident reports a day. Each of these reports is handwritten on a two-page form by the reporting officer. Any additional pages are attached. Before faxes, when reports were filed, they were hand-carried to a central location where they were photocopied then distributed by hand to twelve locations. This process took about four hours from time of filing to time of distribution.

Responding to the call for faster service, the department sought to automate this process of report distribution. They wanted something that would not require significant retraining on the part of their officers, yet that would provide much faster turnaround. The solution was a specially programmed FAXBLASTER multiline fax broadcast machine set up to act as a relay.

Now, when filing an incident report, an officer simply uses a fax machine in any station (every station was already equipped with a fax) to transmit the report to the FAXBLASTER in the operations center. Upon receiving the incident report, the machine immediately begins retransmitting it to a preprogrammed list of other fax machines.

The relay is set up with four lines, and is specially programmed for this application. One line is dedicated to inbound faxes to assure high availability to

OFFICER WRITES REPORT
AND TRANSMITS IT TO
BROADCAST RELAY MACHINE

SPECIALLY PROGRAMMED
MACHINE RECEIVES REPORT
AND REBROADCASTS TO
STORED LIST AUTOMATICALLY

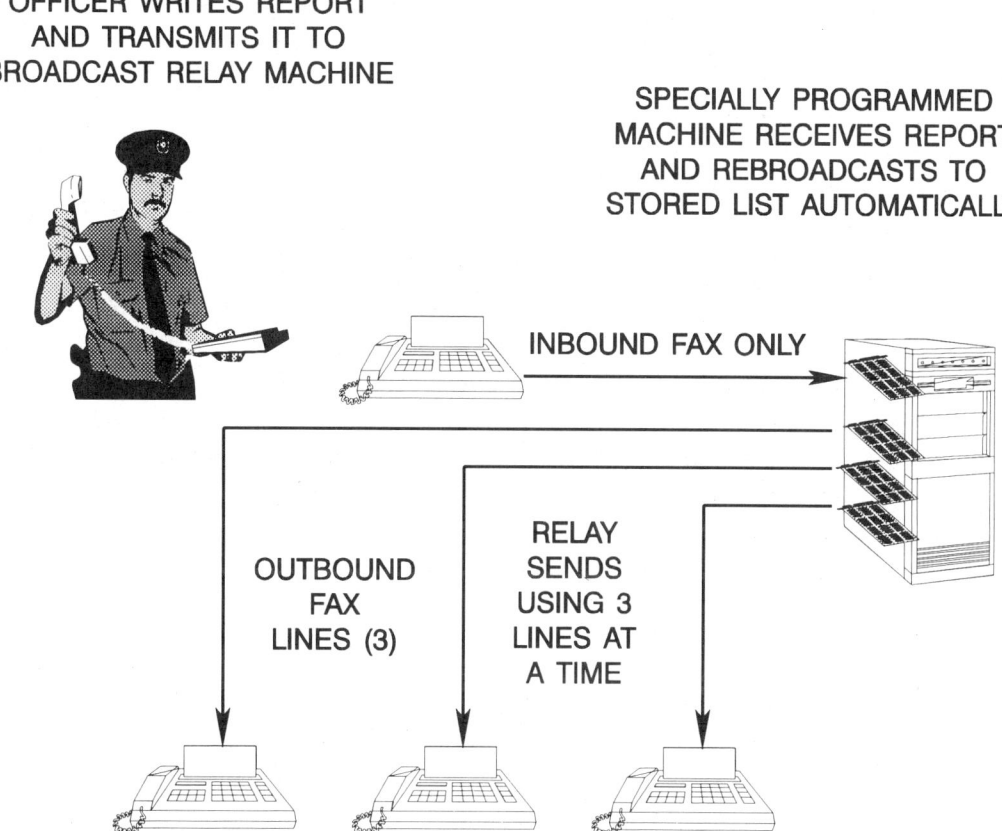

INBOUND FAX ONLY

OUTBOUND
FAX
LINES (3)

RELAY
SENDS
USING 3
LINES AT
A TIME

DESTINATION FAX MACHINES
LOCATED IN CITY OFFICES SUCH AS
DISTRICT ATTORNEY, CHIEF'S OFFICE, ETC.

Figure 3-17: Programmed Broadcast/Relay System

officers who need to transmit reports. This keeps the cops on the street and not waiting in line for the fax. Three lines were configured to handle the relaying of the received reports (see Figure 3-17). Each report, usually two or three pages long, takes about two minutes to transmit. Total transmission time for relaying a two-minute report to twelve locations using three lines is eight minutes. What an improvement over four hours!

Retransmissions happen automatically if the receivers are busy. If several reports arrive at the relay quickly in succession, they are queued up and go out in the order received. The relay requires no manual intervention and is fully automatic. It keeps logs of faxes received and failed transmissions, should they happen, for later operator examination. At the time a transmission or reception error occurs the machine sounds an audible alarm, logs the error, and continues operating. It would be possible to have it send a fax indicating the error but this was not deemed necessary. The automatic relay function can be stopped and restarted at any time so the machine can be used as a normal fax broadcasting system.

Some Do's and Don'ts for Fax Broadcasters

Junk Fax

A ruling by the Federal Communications Commission (FCC) effective December 20, 1992, bans the transmission of unsolicited faxes. Faxes sent without a prior established relationship between sender and receiver will be subject to fines.

Shared Phone Lines

Some fax numbers are actually shared phone lines. This is increasingly the case with home offices. Transmitting a fax to a home office number in the middle of the night can result in an irritated customer. Ask when is acceptable to send broadcasts.

Cover Pages or Not?

Don't send a cover page when it's not necessary. They waste paper and cost you transmission time. The *BioWorld Today* solution to addressing is elegant and effective. If you must send a cover page, consider sending a half page instead of a full page, or even better, use the new stick-on fax labels.

Contact Information

If you are using a cover page, always include your contact information including a telephone as well as fax number where you can be reached should there be a problem with the transmission. Make sure this information is displayed clearly on your fax. Many letterheads use very small, condensed or light fonts for the address and phone numbers below the name or logo. When such a letterhead is adapted for use as fax cover page it is unreadable after transmission. Thus, for your cover page, use a sans serif font of at least 12 points size. We will cover this in more detail in Chapter 5.

Proof Your Documents

Proofread your documents by actually faxing them to a machine to see how they print on paper. If you are using a fax machine, transmit your document to another fax machine for a proof. Don't use the COPY feature, as its results may look significantly better than an actual transmission. Sometimes the COPY feature takes advantage of special gray-scale features, higher resolution, or other proprietary features that enhance copy appearance but will not work for a transmission. You may be surprised to see how the appearance of documents can change after they have been transmitted.

Clean Your Machine

This is important if you are using a fax machine to broadcast faxes or to load a broadcast system. Dirt in the form of ink blobs, lint, dust, and other sticky goo commonly found on documents can get stuck on the scan head of a fax machine.

This causes streaks to transmit vertically down the length of the page. These are not only annoying but they can completely obliterate necessary information on your documents. The COPY feature of a fax machine will reveal this problem if it exists. Copy a blank piece of paper and see if the copy comes out blank. A clean fax is a happy fax.

Set your TTI or CSID.

The TTI, or Transmit Terminal Identifier, sometimes called the CSID, or Terminal ID, is an alphabetical or numeric string that is exchanged between sending and receiving machines during transmission. This number is sometimes printed at the top of each page, and also shows up in the send and receive logs. Set it and keep it up to date as a courtesy to those who receive your faxes. People who use fax modems as receivers will especially appreciate this, as it gives them some idea of who sent a fax before it is displayed on-screen.

Use Standard Mode

Standard mode transmits in half the time it takes to send fine mode, and is acceptable for nearly all purposes. Use fine only mode if you absolutely must.

Busy Machines

If you are sending a long transmission, you're tying up the recipient's fax machine. This can be very annoying if done at a time when the recipient is likely to want to use the machine for critical transmissions. A fax modem or broadcasting system is an automatic tool that makes life much easier for you, the sender. You don't have to wait for retries on busy lines, and your transmissions are queued so you do not need to wait for a line to become free. Recipients may not be in a position of such convenience. Try to be considerate when sending broadcasts, and time them for when the receiver is less likely to have priority traffic.

Turn on ECM

Not all recipients' fax machines will have ECM (error correction mode). But those who do will appreciate your using it. Transmissions can be affected by errors such as scan line drop-out (see Chapter 5) and ECM can help avoid this. Receivers that do not have ECM will not notice any difference. If have it, you should use it.

Follow-up Failed Transmissions

Before merely trying to resend a fax, first find out why it failed. If the transmission failed due a wrong number, some poor person who is called repeatedly may become very annoyed. Think of how you would feel if someone tried twenty times to send a midnight fax to your home phone number.

CHAPTER 4

FAX-ON-DEMAND

On-demand fax, sometimes called fax-on-demand (FOD) or fax response, is the newest and perhaps fastest growing category of fax products and services today. On-demand fax is where the recipient requests a document and it is delivered by fax—instantly and automatically. In Chapter 2 we saw some examples of fax-on-demand in actual applications. This new type of fax was probably first available in 1989, and by this writing literally dozens of products and services exist. The benefits of on-demand fax are clear. People who need printed information can now request and receive it virtually instantly and at any time of day. And the people who would normally have to handle these routine requests for information are now free to handle other tasks.

How It Works

A fax-on-demand system, sometimes called a *fax server* or *fax response system*, is basically a computer system with one or more fax modems (for instance, an Intel SatisFAXtion or a Gammafax CP) and typically equipped with one or more lines of voice processing, using cards such as the Dialogic D40. Voice processing cards permit computers to answer a phone line, play back stored voice recordings, and understand (decode) touch-tone input made by a caller. The entire system is coordinated by special software designed for this purpose. A fax response system typically waits for a request to come in, usually via a phone call, and then responds by sending a fax.

**2: FOD SYSTEM USES
VOICE BOARD TO
PLAY RECORDINGS
AND DECODE TOUCH-
TONE REQUESTS**

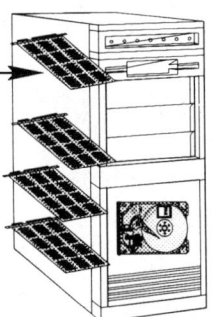

**1: CALLER USES TOUCH-TONE
PHONE TO CALL FOD SYSTEM,
LISTEN TO RECORDING AND
SELECT DOCUMENTS**

**3: AFTER MAKING REQUEST,
CALLER HANGS UP**

**4: FOD SYSTEM GETS
STORED DOCUMENTS
AND PREPARES FOR
FAX TRANSMISSION**

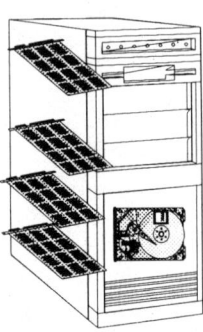

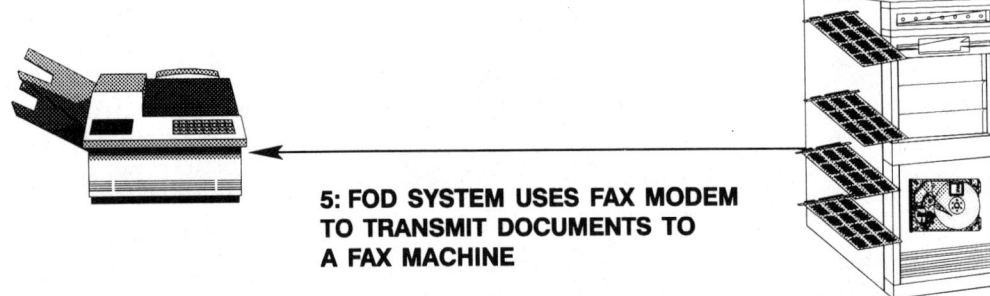

**5: FOD SYSTEM USES FAX MODEM
TO TRANSMIT DOCUMENTS TO
A FAX MACHINE**

Figure 4-1: Two-Call Fax Response

1: CALLER USES TOUCH-TONE PHONE ON SAME LINE AS FAX MACHINE TO CALL FOD SYSTEM, LISTEN TO RECORDINGS AND SELECT DOCUMENTS

2: FOD SYSTEM USES EITHER VOICE BOARD LINKED TO FAX BOARD OR SPECIAL COMBO VOICE-AND-FAX BOARD TO PLAY RECORDINGS AND DECODE TOUCH-TONE REQUESTS

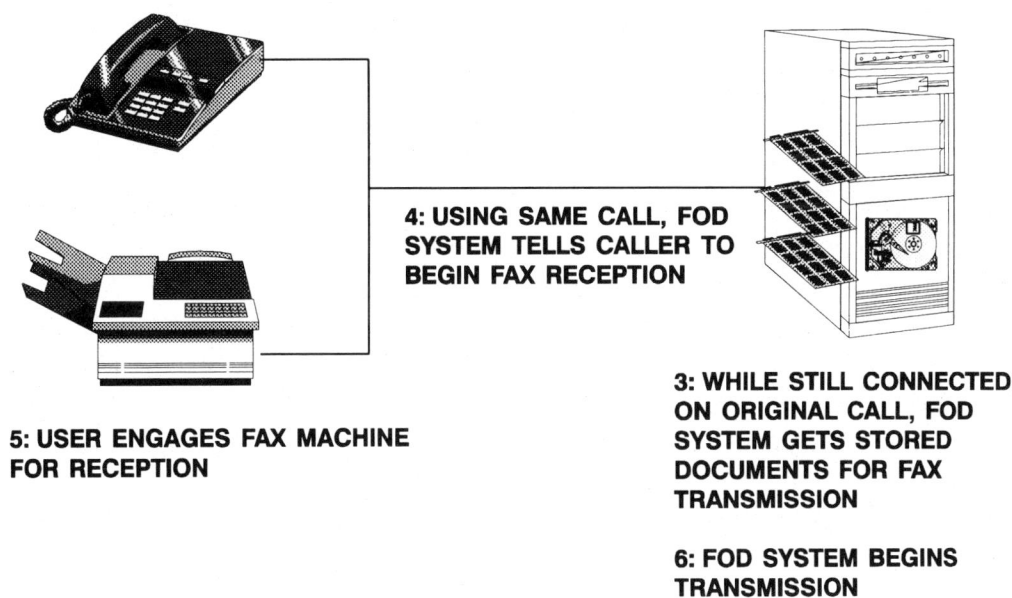

4: USING SAME CALL, FOD SYSTEM TELLS CALLER TO BEGIN FAX RECEPTION

5: USER ENGAGES FAX MACHINE FOR RECEPTION

3: WHILE STILL CONNECTED ON ORIGINAL CALL, FOD SYSTEM GETS STORED DOCUMENTS FOR FAX TRANSMISSION

6: FOD SYSTEM BEGINS TRANSMISSION

Figure 4-2: One-Call Fax Response

Using an on-demand fax server consists of two simple steps: 1) making a request, and 2) receiving material. When a request is made, a fax-on-demand system receives it, figures out what material to send, and then fulfills that request by fax. Two methods of reception are possible: *two-call* interaction, where the caller makes a request on an inbound call and receives a transmission via a second return call (see Figure 4-1), and *one-call*, where both the request and the transmission of fulfillment materials are made on one inbound call (see Figure 4-2).

A fax response system can either contain stored material or obtain it from an outside source. Systems that store information generally retain them as fax images, text files (ASCII), or graphic format files such as TIF, PCX, or DCX. Information can be placed into such systems by uploading via fax transmission, or some form of data transfer such as removable diskette, keyboard entry, or serial port communication. Systems that obtain information from an outside source generally are integrated with a database, either in the same machine or over a network.

The most popular method for taking requests and other caller input is via touch tones in response to voice prompts. A few unusual systems permit requests by fax transmission using optical mark recognition (OMR) or optical character recognition (OCR), although this capability is limited in its reliability. Figures 4-3 and 4-4 show two examples of OCR/OMR forms used by FaxBACK InForms and Xerox PaperWorks respectively. Recorded voice prompts usually lead callers down a "menu" of choices until final selection of material is made. This is the familiar "press 1 for this and 2 for that" interaction. After a selection is made, some systems permit additional selections while others immediately go to fulfillment.

Fulfillment either requires that the caller to hang up and receive documents by return call, or engage the receiving fax machine on the same line for immediate reception.

A typical fax-on-demand system is made up of a computer with disk storage, a

 **FaxBack InForms**™

Document # 1015

1 Please write in your name and any other pertinent information that will insure the proper routing of your information.

Here is the information that you requested from
FaxBack InForms™...

The following information was requested by the party listed below. Please forward this to them immediately.

Name: _____

FaxBack and FaxBack InForms are registered trademarks. All Rights reserved. Copyright © 1992.
FaxBack, Inc. is not responsible for entries made in the above area of this cover sheet.

Using FaxBack InForms is easy!

Simply complete steps **1** - **4** and the information that you requested will be on your desk in minutes.

If you need additional document order forms, feel free to photo copy this form or, simply call the number listed below from the handset of your fax machine. Just press the "START" or "SEND" button when you hear the fax tone.

Sample Telephone Number

2 You may select up to five of the documents listed below:

3 Please enter the fax number (including area code) of where to send the documents.

DESCRIPTION	PAGES
○ An Overview of FaxBack InForms	2
○ FaxBack InForms Configuration Options	1
○ FaxBack InForms Auto Update Feature	1
○ FaxBack InForms Auto Update Form	1
○ FaxBack InForms Feature Set	3
○ FaxBack InForms Invites Your Feedback	1
○ FaxBack InForms Press Release	2
○ Example of a Credit Card Application	1
○ FaxBack InForms USA Order Form *(This Form)*	1
○ FaxBack InForms International Order Form	1

Return Fax Number

FaxBack Informs will automatically deliver the information you requested within the next few minutes. There is a possibility of other document requests being queued before your's. If that should happen, there may be a short delay. Thank you for your interest in FaxBack Informs.

Figure 4-3: FaxBACK InForms OCR/OMR Form

PaperWorks

Starter Form

Please do not write above this line

SECURITY CODE	A B C D E F G H I J K L M N O P Q R S T U V W X Y Z

RETRIEVE

☐ In Basket (all items)	☐ In Basket (new mail)	☐ Activity Log

LIST CONTENTS

☐ In Basket (all items)	☐ In Basket (new mail)	☐ Document Index
☐ Recipient Index		

Into the following form section(s):

☐ Retrieve	☐ Send	☐ Delete

RETURN ADDRESS

A Return Address fax number is required when you want to Retrieve or List Contents.

Check the fax number for where items should be sent.

For U.S. calls, check numbers for area code and local number. For international calls, check numbers for country, city, and local phone number. Do not skip columns.

Mark an X in the box for each number.
Mark only one box per column.
Example: (415)283-6870

Peter Lafreniere: 03/22/92

Figure 4-4: Xerox PaperWorks OCR/OMR Form

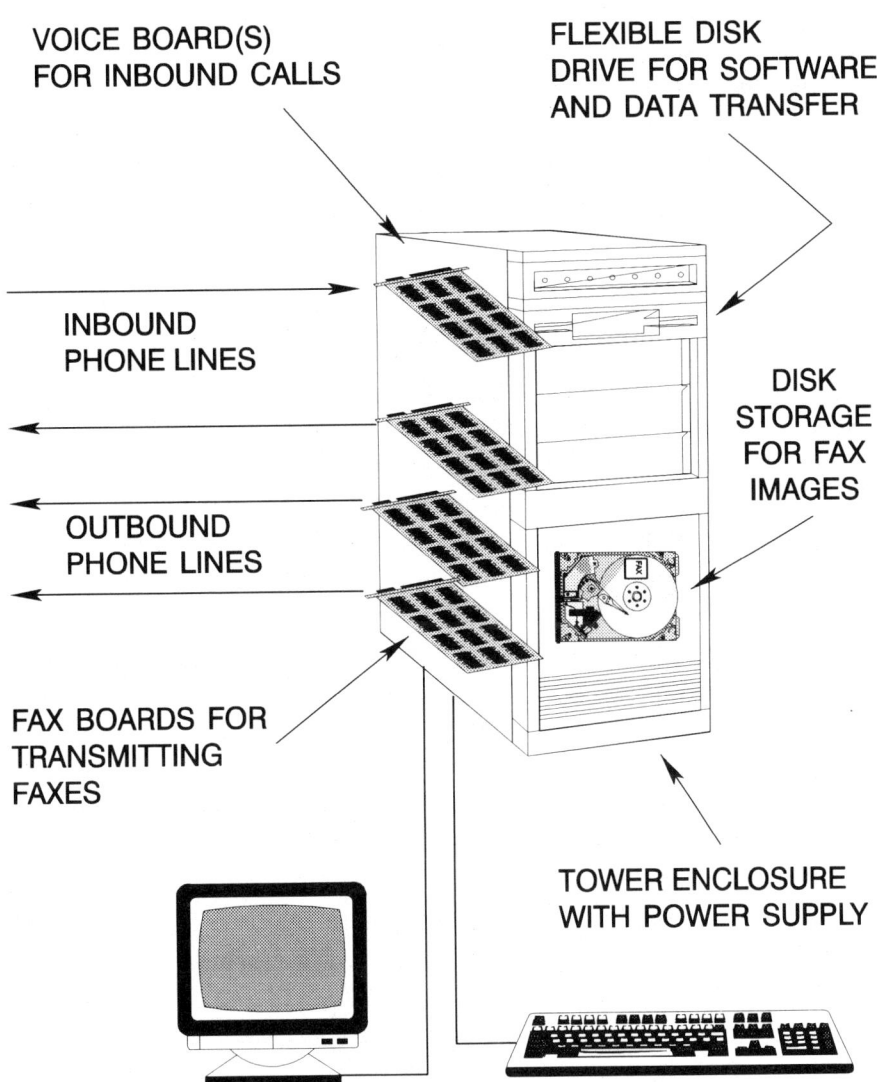

VOICE BOARD(S)
FOR INBOUND CALLS

FLEXIBLE DISK
DRIVE FOR SOFTWARE
AND DATA TRANSFER

INBOUND
PHONE LINES

DISK
STORAGE
FOR FAX
IMAGES

OUTBOUND
PHONE LINES

FAX BOARDS FOR
TRANSMITTING
FAXES

TOWER ENCLOSURE
WITH POWER SUPPLY

MONITOR AND KEYBOARD

Figure 4-5: Anatomy of a Fax-on-demand System

voice processing board, and a fax board (see Figure 4-5). Some systems may have keyboards and monitors for use by an operator, and others may have multiple voice boards and fax modems. A few systems use boards that can perform both voice and fax functions. The vast majority of these systems are designed to connect to an ordinary phone line using a modular connector. Some systems are able to handle ground start phone lines as well as the regular loop start lines. A few that are designed for high-capacity applications can directly connect to a digital T-1 line, which handles twenty-four calls simultaneously.

Special software that runs on the main board of the computer controls the operation of the fax-on-demand system. The entire market of prepackaged fax response systems seems to be dominated by products based on the IBM PC AT-type personal computer. This software controls when voice recordings are played or recorded and when faxes are sent or received. It may also control a keyboard and screen interface for an operator. The voice boards are used to play and record voice prompts and to detect touch-tone input from the caller. The fax boards are used to send and receive fax documents. The fax documents are stored as files on the disk drive, and may be in one of many formats. The number of documents that can be stored is in direct proportion to the size of the document and the amount of disk space.

Most systems use a document numbering scheme to allow callers to select documents. For example, a document might have a three- or four-digit number that the caller would enter in response to a prompt to order it. A few systems rely on scripted voice prompts to allow document selection. For example, a series of prompts might ask the caller to "Press 1 for price information and 2 for product information." The former has the advantage of more flexibility and ease of use for a larger number of documents. The latter has the advantage of simplicity for fewer documents, typically less than a handful.

Performance, Selection, and Usage

A variety of factors affect the performance and selection of how a fax-on-demand system might be applied to solve any given problem. Because this type of product is relatively new, vendors tend to use different names for the same features, often adding to the confusion by trademarking them. This section should provide a basic understanding of generic fax-on-demand features, why each might be important to you, and how they work.

One-call versus Two-call

Early equipment was predominantly two-call, largely due to technical limitations. But this has been changing quite quickly. The trend is toward systems that can do both one- and two-call. Which method you select should be guided by the nature of your intended application. Here are some factors to consider before choosing one- or two-call.

Instant, guaranteed delivery This is the strongest reason to prefer one-call over two-call. This is the key difference between the two methods. Only the one-call method can assure that fulfillment will happen instantly while the caller is standing in front of the fax machine. It guarantees that if an inbound call goes through, the line will also be available for the return fax.

With two-call operation, return fax of documents can be delayed if inbound requests arrive too rapidly, causing a queue of outbound transmissions to form. Return transmissions can also be delayed or obstructed completely if the line for the receiving machine is engaged due to heavy use or out of order. One-call operation avoids these pitfalls because the caller receives fulfillment by holding open the line used for the inbound request and engaging manual reception. The drawback to one-call is that the caller is required to originate the call from a handset or other instrument attached to a fax machine. Two-call allows requests to be made from practically any touch-tone telephone.

Routing No reasonable method has yet been devised to allow recipient's name to be placed on a cover page when using fax response without a preassigned identification number or other means of identification. Several interesting approaches have been tried, and some are even acceptable for the technically adept. Most makers of two-call products have attempted to mitigate this problem by asking the requester to input either a telephone number or extension number at the time of the request. This number is then used to generate a cover page that asks an operator to deliver the fax to the person at that number. A few innovative developers provide ways a caller can encode alphabetical characters, but these are all still too difficult for the vast majority of callers. An interesting example is FaxBACK, Inc., which uses a Morse-code-and-keypad-position approach. To enter "PHIL," you would type 7, pause, 44, pause, 444, pause, 555. The way a caller is supposed to remember this is that on a touch-tone keypad, the letter "P" is the first letter on the "7" key, the letter "H" is the second letter on the "4" key, and so forth. During each pause the machine reads back a letter and asks you to confirm it.

The inability to put a name on a cover sheet of a transmission destined for a shared machine often results in problems with delivery. The one-call method avoids this problem entirely and has the additional benefit of eliminating the cover page.

Avoidance of Outbound Telephone Charges This is the other reason most often cited for using one-call over two-call. With the one-call method the burden of telephone charges can be imposed entirely on the caller. This is important for organizations seeking to use fax response as a way to reduce costs as well as augment service. Many government agencies have taken to using this method based on the belief that the increased level of service provided warrants the cost of a telephone call on the part of the users of such a service.

Two-call operation is advantageous when you anticipate that the majority of callers will be inconvenienced by having to originate their calls from a fax machine. This is the primary reason given for using two-call over one-call. Some claim it is also simpler, since the caller does not need to know how to operate the fax machine in manual mode. These claims have some validity, and for situations involving promotional materials, two-call tends to be favored. There are, however, other reasons to use two-call over one-call.

Forwarding of Stored Fax Documents is probably the greatest reason besides economics for the use of two-call. Forwarding is where the requester calls a fax server and directs material to the fax of another party. The typical application is sales, where the salesperson makes a request on behalf of a prospect or customer. If users for a fax-forwarding-on-demand application are easily identified and trained, then the routing problem may be solved by the use of an alpha encoding method to place the recipient's name on a cover page.

Acquisition of Fax Numbers is now questionable due to a December 20, 1992 FCC ruling that bans "junk fax." This law makes it a crime to send an unsolicited fax transmission to any person with whom the sender does not have a "prior relationship." It is unclear whether requesting a document from a fax server constitutes the basis for a "prior relationship" between the operator of the fax-on-demand system and the requester. Another possible use of such acquired fax numbers would be to look up the subscriber names using a reverse directory and then engage them with standard direct marketing techniques. Experience shows that capturing such numbers has generally been unproductive.

Security is an sometimes cited as a reason for using two-call, since some fax servers have extensive programming features and can be set up to only forward certain materials based on a password or other identification. It could be argued however, that callback is inherently less secure than one-call. With one-call, in addition to the possible use of a password, you also have the assurance that the material is going directly to the caller, who is in front of the machine at the time of transmission. This decreases the chance that a fax operator may spy on a transmission or that it may get misplaced or delivered to the wrong person.

Time Delay may be a very valid reason for employing the two-call method over the one-call method. If the requested materials are produced on a schedule and not available continuously, or if substantial processing is required to fulfill a request (for example, the generation of a database report), then two-call may prove much more convenient for the caller.

Executive Use There is a common belief that executives would never go to a fax machine to use a one-call fax server, and therefore when a service is intended for executive-type callers, it should be two-call. This is probably not a real factor, since the most important consideration is the content of the server and not its access method. A *real* executive would probably have an underling do the work anyhow.

Availability of Touch-tone Equipment Touch-tone service is very common in most of the U.S. but this is not true of many other places such as England. If your calls are often international, you may need to consider this. Nearly all fax machines have touch-tone capability, but in places where callers do not have touch-tone service the telephones will probably not have touch-tone capability. Thus one of the benefits of two-call, the ability to call from a telephone and have documents sent to a fax, may be eliminated for lack of touch-tone telephones. Callers may need to use their fax instead, because it may be the only device that can has the touch-tone feature. Thus, if touch-tone telephones are not available, it may be necessary to instruct callers to call from their fax machines and switch them temporarily to touch-tone operation for the duration of the call to your fax response system.

Use With Pay-per-call Services Pay per call services, also known as 900 and 976 numbers in the U.S. or 0898 numbers in England are used to charge callers either a flat rate or on a per-time basis for a call. One-call service permits time-sensitive billing for information, where two-call allows the simple implementation of per document billing.

Storage Capacity

Documents stored on fax servers usually reside on an internal computer disk drive. Drives come in various sizes, and size is one of the factors governing how many documents a fax server can store. Some makers offer varying sizes of storage capacity. Capacity is sometimes described in terms of "pages," but the definition of a page can be ambiguous. A better gauge is the amount of usable storage expressed in megabytes (Mb). Given that number, you can estimate that each megabyte of storage will hold about twenty pages of typical documents (no pictures, screens, or very dense type). To figure this more closely using actual documents for a specific application, you will need to obtain an average document size based on actual measured disk space consumption. One way to do this is to fax your documents to a fax modem equipped computer or broadcast system and then measure the size of the corresponding files. If you have only a fax machine available, you can apply the following method:

1. Transmit your document to another fax machine and measure the time required.

2. Apply the following formula to compute estimated document storage size:

   ```
   1.21[(t-13)-(2n)] = size in kilobytes
   ```

   where n is the number of pages and t is the transmission time in seconds.

3. Divide the above size in kilobytes into 1,000; this gives the number of such documents per Mb of storage.

4. Multiply the number from (3) times the total storage in Mb and that gives total pages of storage for your typical document.

This formula is accurate to about 10 percent for documents twenty seconds and longer in transmission time. Transmission speed must be at 9,600 bps.

Call Handling Capacity

The cost of fax servers usually varies with the number of lines a system can handle. A fax server should have enough lines to handle the needs of its intended application. Required call handling capacity depends on:

1)   The time it takes to make a request and transmit a fax in response.

2)   The degree to which the system operator wishes to avoid call blocking (busy signals on inbound calls).

3)   The need to avoid delays in outgoing fax transmissions.

Processing time for most stored information systems is generally negligible by comparison with transmission time.

For one-call systems you can estimate the average call hold time (the total time a line is busy due to a single instance of a caller using the system) by adding up the average estimated request and response times. For two-call systems you should consider these two separately, as call hold times are likely to be considerably different (usually shorter) for requests than for responses. The result is that the number of inbound and outbound lines will not be equal. For example, let's say the average call hold time on an inbound request for a two-call system is two minutes and the average transmitted response is four minutes. Without doing much figuring it is possible to see that with equal numbers of inbound and outbound lines, requests could pile up much faster than responses could get out.

As a first order estimation, you can use Table 4-1 to approximate how many lines an application might need:

	AVERAGE CALL HOLD TIME (DURATION) IN MINUTES			
LINES	1	2	3	4
1	30	15	10	7
2	90	45	30	22
3	150	75	50	37
4	210	105	70	52

Table 4-1: Calls Per Hour Capacity versus Lines and Average Call Hold Time

When considering how much capacity you will need, consider first the acceptability of having calls blocked or responses delayed. Many applications can tolerate substantial queuing of outbound responses or blocking of inbound requests, thus reducing the need for additional equipment expense. For other applications, availability is critical. Try to characterize use in advance or provide for smooth increases in capacity should it be required. A single line system running with an average call hold time of 3 minutes can handle 80 calls in a business day. Although this is more than adequate for many applications, the overwhelming tendency is to purchase excess capacity.

Voice Menus

Some systems come with extensive voice menu structuring and programming capabilities. Others come with only the ability to record a greeting. The tendency with voice menus is to make them too long-winded with too many layers. *Keep voice menus brief.* This cannot be overstated enough. The rate at which calls are abandoned is directly proportional to the length and depth of voice menus. If you use a system with voice menu programming, keep the prompts brief and provide an easy means for callers to obtain help by fax or voice at any time. For example, you might tell callers they can "get help at any time by pressing the "star" button and then offer context-sensitive instructions by voice or fax. A picture is worth a thousand words, and the same could be said for sending a fax versus listening to voice menus. A well-structured application will help new callers by offering instructions and a directory of documents by fax rather than requiring them to wade through voice menus.

Callers to fax response systems can generally be split into two camps: one-time or infrequent callers who will need more assistance and will be willing to take more time with voice instructions, and repeat callers who habitually will use a system to obtain information of recurring interest to them. Repeat callers display a lower tolerance for longer voice menus but are willing to invest more time in learning shortcuts or how to work through a terse set of instructions.

Generally, the population of users for a given application will consist predominantly of one type or the other.

When you design an application, the expected usage patterns of the system should influence the choice of system features and the design of voice menus.

Call Restriction

Although you not may think your application requires any security consider this question: How do you keep someone from making your fax server repeatedly dial a 900 number? What do you do if you want to provide two-call service for local calls only? The answer is a feature called *call restriction,* which lets you restrict calls using various criteria such as area code, prefix, and time of day.

A system using two-call fax response without call restriction is a potentially large telephone bill waiting for you. Toll calls to far-away places can be very expensive and you should assume that hackers, twits, and other troublemakers will try to abuse your system for little more reason than perverse entertainment.

Access Control

If you intend to run a "closed" system where only certain callers are allowed access to your fax-on-demand information then carefully examine the scheme for providing different types of access. Is it easy to reset a forgotten password, add or delete a user, or implement different restrictions for different callers? Some systems do not come with any keyboard or monitor, and therefore can be programmed only by using touch-tones from a telephone. Others operate just like a computer, with menu-like interfaces. Restrictions can be imposed on an individual basis in most cases by requiring a user identification number or password to request certain information. But your application may need multiple levels or restrictions based on groups, time of day, telephone line used, or other variables such as number of times an ID or password has been used.

Carefully consider the needs of your application in this area, as security schemes can vary widely among systems.

Reports

Most operators of fax-on-demand systems want to know how much their system is used. Some want to know who is using it. Others just want to acquire the fax numbers of their callers. And finally, a few will need to make reports to management based on information obtained from the machine.

Fax response systems offer a wide variety of reports ranging from the terse to the data-intensive. At a minimum you should be able to tell how many calls came in and what information was faxed out. Other options to consider are:

- *Dialed fax number.* For two-call setups, the reports show the fax numbers dialed on the outbound calls.

- *Transaction logs* show when calls came in and went out, who called (by user ID), what fax number was dialed back, or if the call was one-call, what documents were selected and in what order; and if the transmission completed. Transaction logs can also show at what point in a voice menu calls are abandoned.

- *Usage statistics* summarize data such as the total number of calls, total number of times each document was selected, last time a report was generated, and the last time the counters for these statistics were reset. Knowing what information is popular and how many calls are handled can help to guide your decisions about what to place on the system and whether more lines may be necessary.

Extensibility and Interoperability

Fax-on-demand functions do not need to stand alone. They can be integrated into other equipment and systems such as PBX switches, voice mail systems, databases, local area networks, and mainframes. Some fax response systems and software packages are designed to support interoperability with popular databases such as dBase, Paradox, and Foxpro, or operate over local area networks or with mainframe applications. Many PBX vendors that sell voice mail as a feature have added fax-on-demand to their existing products. Voice mail manufacturers have also added fax-on-demand ability to their products. A few telecommunications industry watchers believe that the integration of voice, electronic mail and fax will happen in the near future, allowing all these functions to be integrated through personal computers or workstations. Should this come to pass, fax-on-demand will become incredibly commonplace and will probably be extended to include voice and data as well. If you already have a voice mail system or a PBX, you may want to begin by inquiring about upgrading it. Adding a feature to your existing switch or voice mail may cost less and integrate better with your current operations. On the other hand, getting mixed up with more departments may be the last thing you want to do—which may be the reason there will always be room for standalone products!

Fax Mail

Fax mail can be a handy feature and seems to be a natural extension to a fax-on-demand system. Fax mail lets callers send faxes to a mailbox on a fax server for storage and later retrieval by the owner of the mailbox. Some systems that have a fax mail as a feature will also allow automatic forwarding of stored faxes, often at a specific time, such as when telephone rates are lower. An interesting augmentation to fax mail is pager alert. This is where the fax server calls your pager number to notify you that fax mail has arrived.

Credit Card Processing

If you intend to sell information using a fax-on-demand system, then you may want the ability to process credit card transactions automatically. A fax server with a credit card processing feature can take the credit card number and expiration date from a caller over the phone (using touch-tone input) and immediately process a charge to that account. Most work like this: After receiving the credit card number and expiration date, the fax server tells the caller to hold and dials a credit card processing service such as BT Tymnet or Modular Data Systems. The server emulates one of the common point of sale draft capture terminals typically used by retailers. When the authorization is received, the caller is notified and the fax is sent. Your bank account is electronically credited.

Voice Mail

Many fax-on-demand products incorporate a voice mail feature, just as many voice mail systems are adding fax response as a feature. In a fax server, the typical purpose of voice mail is to allow the caller to leave a message typically containing the caller's phone or fax number. In marketing applications, for example, this is used as a means of capturing sales leads. If your needs for voice messaging are extensive, you may want to consider a commercial voice mail system with a fax response add-on feature. If your main interest is dispensing faxes, however, your best bet would still be a fax-on-demand server.

Training

System Operators

The larger and more complicated the fax-on-demand system, the more time spent training people to operate and maintain it. If a fax response system is to save time and effort, it should be kept as simple as possible, at least at the start.

System features can become so numerous and complicated that nobody ever uses them. Consider your cost of learning a new system during the selection process.

Callers

Don't pick anything that requires a lot of caller training. All the instructions a caller should need prior to calling a fax-on-demand system should be contained in a single brief line such as "Call this number from your fax" for a one-call system or "Call this number from a touch-tone phone" for a two-call system. The voice response system at my credit union requires a reference card to operate it. This is the wrong approach.

Input Options

There are several ways to get information into a fax response system. Faxing, scanning, transfer by diskette or tape, and keyboard entry are all possible. If you are going to create digital direct fax documents, then input by faxing direct or by media transfer are the only real options. If most of your information is in print but not on a computer or desktop publishing system, then faxing or scanning is the way to go. Most fax-on-demand applications are motivated by a repeat demand for certain kinds of information. How much information, how often it changes, and in what form it is generated or stored will guide what input method seems best. Very rapidly changing information, such as weather maps or stock prices, may require direct machine-to-machine integration. Keyboard entry may be acceptable if the amount of information is small and text is the best way to store it.

Fax-on-Demand Equipment—What's Out There

A wide variety of fax-on-demand products are now available, including custom-built equipment, complete systems, computer add-on boards, and software. The range of what's available goes from about $500 for a do-it-yourself kit of boards and software to custom-built systems capable of doing almost anything with fax-on-demand you can imagine. There are many choices, and in this section we will examine various kinds of FOD products and look at specific examples of each. After reading this and the previous section, you should be knowledgeable enough to specify and select FOD equipment.

Packaged Fax-on-Demand Systems

By the end of 1992 the number and variety of packaged systems on the market had greatly expanded compared to what was available only a year before. Prices for systems can range from a semi-customized FaxBACK Rack from FaxBACK, Inc., selling for $75,000 to $100,000, to the mid priced FlashFAX from Brooktrout Technology ($5,000 to $7,000), to the downright inexpensive ($1,995) MessagePost from DBC Associates. The Appendix lists vendors of fax response products.

Players in this market seem to be coming from two directions but converging on the same place. Suppliers of fax-on-demand equipment started with the assumption that the goal is to deliver a document by fax. Voice response and related features were incorporated only to the extent that they helped enhance the delivery of fax documents or met a specific market demand. Vendors of more established voice mail and interactive voice response (IVR) products viewed fax-on-demand as an enhancement to their existing systems which were mainly meant to answer the phone, interact with callers, take messages, and play recorded information.

These two approaches will somewhat determine what products are better suited to your needs. For example, the makers of FOD systems typically have more features that make handling fax documents easier. Many provide some form of format conversion, viewing, editing, loading or unloading, updating, and interfacing with other systems such as databases. The fax is viewed as the product, so more attention is paid to how it is managed. With voice systems, there is usually a lot more sophistication on the voice side. Typically, such systems have full voice mail features, including integration with popular PBX equipment. Since many see fax as a "mail" feature and not a delivery feature, emphasis may be placed on keeping received faxes private.

Example fax-on-demand systems are:

Brooktrout FlashFax. One example of a typical packaged fax-on-demand system is the Brooktrout FlashFax. This system, which sells starting at $7,995, uses a two lines; one voice line for handling inbound requests via touch-tone phone, and a different fax line for transmitting the requested documents. FlashFax comes with an 80 megabyte drive, giving it the ability to store about 1600 typical pages of information. The operator loads documents and controls the operation of the FlashFax by using a separate fax machine and the touch-tone key pad of the handset or attached telephone. The FlashFax is also capable of recording voice messages from callers. This feature can be useful in gathering information such as sales leads and trouble reports. FlashFax handles all requests in two calls; one inbound for requests, and a separate one for transmitting fax documents. The FlashFax uses the Brooktrout TR112 2-channel Fax Card and the TR200 2-channel Voice Card.

FaxBACK, Inc., FaxBACK Rack. For the technically ambitious or those desiring maximum features, there is the FaxBACK Rack from FaxBACK, Inc. Each rack can handle up to forty-eight inbound voice lines and twenty-four outbound fax lines. The system runs on a Novell local area network and is controlled from an operator's console and file server. Combined with the Master FaxBACK feature package, this large system can also provide up to 999 different programs for callers based on inbound line, caller provided extension number, or direct inward dialing (DID) number. The system can be custom interfaced with a host computer to provide database access via fax-on-demand. A single rack can work in multiple languages simultaneously. Additional features allow the processing of credit cards, fax mail, voice mail, voice request capture, remote updating of voice prompts and fax documents, live operator transfer, and a wide range of access control over callers and destination fax numbers. Each system is built-to-order and priced based on capacity and features. FaxBACK uses Dialogic voice cards and Intel SatisFAXtion fax modems.

DBC Associates MessagePost. At the opposite end of the spectrum is the entry-level end of the market represented by the single-line, $1,995 MessagePost from DBC Associates. This simple machine is more like an answering machine than a fax response system. It operates in one of two modes. In *standard mode*, it will store documents and take requests by document number, just like most systems. In its unique *simple mode*, it dispenses a single predetermined document only, avoiding any prompting for document numbers. The MessagePost is ready-to-run and does not need any programming. This one-call system stores up to 800 pages of typical documents on an internal 40 megabyte disk drive. All documents are loaded by calling in from a fax machine, and all control of the system is via touch-tone keypad. The MessagePost uses a proprietary single channel card for both voice and fax.

ABS Systems, Inc., TALX. An example of voice processing systems that have added fax-on-demand features to their products is the TALX voice processing system from ABS Systems, Inc. This self-contained system can operate with any PBX or key phone system and will support from two to sixteen voice lines and a single fax line. It is compatible with Centrex and DID phone lines. It performs auto attendant, voice mail, interactive voice response, as well as fax-on-demand. The system is fully programmable via on-screen menus. TALX systems start at $8,150 for a two-port (one voice, one fax) configuration. ABS uses their own proprietary voice and fax cards.

Hardware and Software Component Products

The variety of roll-your-own boards, software packages, and kits is great. Here is roughly how the market is segmented: Software vendors typically place their value in their programs and sell hardware only as a convenience to their customers. Most software uses the Dialogic D40 or D41 family of voice boards; the few that use drivers for the Rhetorex, Pika, and other voice boards have less market share. Dialogic virtually owns the voice board market, so it is not surprising that most software developers would target that platform. Board makers often supply software with their products, either bundled or separately. A few board makers, such as Dialogic, do not need to provide software to end users because much of the existing software uses their product. Some, such as Intel, provide drivers and other support utilities, but the applications are from third parties. Finally, products can have either a voice or fax orientation. Most of the products in this area are transplants from the world of voice processing that have added fax response to their feature set. Voice processing has been around longer than fax response. We will focus on the software, since it is the real driver of functionality.

The single greatest benefit of doing it yourself is the cost savings. You can put together a system for less than you can purchase one, especially if you have some old 286 or slower 386 machines laying around looking for work. Building a fax response system can be a good way to recycle used computers. Those who are already PC technicians will have no difficulty using any of the commercially available component level products, as they are not much more challenging to install than most system-level DOS-type applications. If you don't have technical knowledge, you are probably better off going with a prepackaged system, which will save you time and headache.

The following covers some of the available products:

Talking Technology, Inc., Fax Mouth The low-cost-of-entry award in the fax-on-demand component market has to go to Talking Technology, Inc., and its Big Mouth and Fax Mouth board and software combination products. For around $500, depending on where you buy, plus the cost of a 286/386-type computer system, you can put together your own single-line FOD system. The Fax Mouth is a fairly typical product that includes features such as number blocking by area code and prefix, day and night variable messages, password protection by document, attendant transfer, and voice message recording. But it comes with a special twist. It provides a simple way to get digital direct faxes into the fax-on-demand system by providing compatibility with BitFax for Windows from Bit Software. It works by using BitFax to create the fax files, which are stored for the on-demand part of the system. As we will discuss in detail in the next chapter, any Windows application that can print can be used to create a digital direct fax document with print utilities such as BitFax.

SpeechSoft, Inc., Speech Master. This software package was first intended for controlling automated call processing systems for applications such as voice mail, auto attendant, IVR, and outcall. Recently, a fax component was added, allowing Speech Master to use any CAS compatible fax modem to provide fax response service. Speech Master supports up to four fax lines in either one-call or two-call modes, and gives the system operator total application flow control over both voice and fax interaction at the programming language level. Other notable features of the product include ability to directly access Novell Btrieve databases, full real-time display of activity on-screen, and capacity of up to twenty-four lines. This program is a full-blown development environment for voice and fax applications.

Kiss Software, Voice-Tree/Fax-Tree. Most voice processing and fax-on-demand software is intended for use by value-added resellers (VARs), not for end users. Voice-Tree/Fax-Tree is marketed for end users and is intended to bring the capabilities of application development to people who are creative end users but not necessarily technical developers. While it is still not quite to the stage where a complete novice can program it, a computer-literate nonprogrammer who knows about voice and fax communications should be able to handle it without too much difficulty.

To use Voice-Tree/Fax-Tree you will need a 386-type computer, a phone line, a voice processing board from Dialogic, Rhetorex, Pika, Linkon, or New Voice, and an Intel SatisFAXtion fax board. Voice boards cost from about $400 for a single-line to $1,400 for a four-line board. Also recommended is QEMM from Quarterdeck, a memory manager that is required if you intend to run multiple lines. Setting up the machine is the most complicated part of the process.

Once the machine is set up, a demonstration program is provided for testing the system, and the package comes with a preprogrammed voice mail application. Programming of your application is done using a large on-screen table. You can use the prerecorded voice files or record your own. The basic software is $995 and the fax add-on feature is $250. A limited edition two-port voice and fax is $499.

Xerox PaperWorks. Perhaps the most unusual of component products is PaperWorks from Xerox Corporation. This product is completely different from the others. PaperWorks runs under Windows and is intended for the PC user as a means of accessing files and faxes remotely using a fax machine and special machine-readable forms. The creators of PaperWorks view the world from the perspective of document management. There is a whole industry dedicated to document and image management, with its own international association and annual trade show. PaperWorks lets you store, send, or retrieve any document stored on your computer, either while you are at your computer, or remotely using a fax machine and its special paper forms. The idea is that your computer is an electronic filing cabinet and the fax is your remote access terminal to that cabinet.

To store a document, you place a special form on top of your document and fax the entire lot to your machine. An example of a special form is shown earlier in this chapter in Figure 4-4. When your computer receives the special form, PaperWorks processes it and puts the document away for later use. You can send a document stored in your computer either to yourself or the fax of another using a similar special form and a fax machine. You just fill out the form and send it in. PaperWorks receives and processes it and then send out the appropriate document. Similarly, you can retrieve any file on your computer that you have decided in advance to make available to remote access through PaperWorks.

PaperWorks comes with some prepared forms and also allows you to custom generate forms that contain functions and information specific to your task. For example, you can create your own checklist of items that list only specific documents. People can then use this form to order those documents, but because they don't have any other forms, they are restricted to ordering only those on the list. Each of the forms contains a special area called *glyph* which helps PaperWorks identify and process the form. The glyph area of a form contains approximately two square inches of small herringbone-like marks.

If you send in a document that is not one of the special forms, then the document is not processed but held for later viewing by the operator. This is similar to how regular fax software operates. PaperWorks can also perform most of the other functions of fax software, such as sending, receiving, and maintaining a phone list.

Does this sound like fax-on-demand? Xerox says to "Look at it as a 1/4 ounce pen computer," but the product works just like fax-on-demand. You can store documents by fax transmission, copy them in from diskette, or create them on the computer itself. Access to the information is controlled by the special forms. Requests are received by fax and returned by fax. Operation of the product takes place in the background and is controlled by a Windows interface program.

One caution should be observed regarding PaperWorks, and this warning is applicable to all programs or services which claim to be able to use OCR/OMR to process fax-on-demand requests. The ability of OCR/OMR technology is very delicate. In order to function correctly, transmissions must be clean and very limited in the amount of skew or stretch. Original forms should be used when possible, and in the worst case, first generation copies. Users must be aware that nearly all these forms contain critical areas that must not be obstructed, obliterated, or altered. Such critical areas allow the recognition software to locate the page, orient it correctly, and find the areas to process. Users must be trained to make clear, distinct, well-bounded marks in check boxes and other data entry areas. Fax transmission equipment must be maintained to a high level of cleanliness, as streaks and other noise due to dirt on scan heads, or nonlinear distortions due to dirt on pinch rollers, will cause failure in forms processing. Finally, high-quality phone lines and connections are a must, as scan line dropout and other transmission-induced problems are another potential cause of failure.

With a high degree of user compliance and a clean transmission environment, packages like PaperWorks should be able to achieve a high rate of success in processing forms. Such a rate, in the 95 to 97 percent range, should be sufficient for most noncritical fax-on-demand applications. In an uncontrolled environment, where users are not trained or operation of the transmission equipment is unrestricted, 70 percent accuracy is regarded by experts to be the theoretical best performance. If you are serious about using this kind of technology, an extensive program of testing under the actual conditions of your application should be performed and evaluated.

In any case, PaperWorks is probably still be worth a look. As an educational or experimental vehicle, it may provide more than enough value to justify its relatively low cost of under $300. And if is not used for fax response, PaperWorks can still function perfectly well as conventional fax software in the Windows environment.

Fax-on-Demand Service Bureaus

Fax-on-demand service bureaus will dispense your documents to callers using their equipment—all for a fee, of course. In addition to providing fax response service, they also provide many other valuable support services, such as creating digital direct faxes from PostScript or other formats, loading and managing fax documents, and even maintaining database information. *Cruising World's* "Another Opinion" fax-on-demand service is handled by *Instant Information* of Boston. Most service bureaus provide both on-demand and broadcast services. While features and services differ somewhat between service organizations, the basics are fairly constant. Here are some points to look for when examining a potential service bureau.

Communications Methods. Nearly any service bureau should be able to support 800, 900, and local calling from almost any carrier.

Billing Methods. Credit card processing, 900 numbers, and usage-based invoicing are all ways you can charge callers or you can be charged by the service provider. Some service bureaus will provide factoring of credit card transactions. This is where the service bureau acts as the merchant that accepts credit cards, obviating the need for you to obtain merchant status with a bank. They charge the caller and then pass the funds, minus a discount fee and a service charge, on to you . If you are using 900 numbers, be sure to check the service bureau policy regarding proof-of-delivery and chargebacks. Chargebacks happen when callers refuse to pay for service obtained via a 900 number, often claiming the service was never received or of poor quality. When a chargeback is allowed, the carriers, which are the collecting agents of the funds from callers, will not collect from the caller and will not pay you, the information provider. But you still get charged for use of the carrier-provided 900 service. There are businesses selling reports via FOD that cost in the hundreds of dollars, and the liberal 900 chargeback rules could present a wide open source of after-sale liability.

Reporting. Reports can provide a wealth of information and you should be sure to ask your service bureau about all possible information you might want now or in the future. Some types of information which may be available are listed below. You may have other needs, depending on the nature of your business.

- Calling numbers, which are the telephone numbers from which calls originated.

- Call hold times showing how long callers were using a specific voice program.

- Abandon rates that tell how often callers hung up without requesting material.

- Transmission logs showing what fax numbers were called, when, what was sent, and if transmission was successful.

- Statistics such as number of total calls, tally counts to show how many times each document was selected, and calls distributed by time of day or place of origin.

Retry Strategy. If you are using a two-call method, then the retry strategy will be important, especially if you are sending long documents. Sometimes calls can be cut off in the middle, and if you are paying for retries, the fax retransmission should be able to reconnect and pick up where it left off rather than resending the entire document. You may also be able to control the retry delay and to select the number of immediate redials and delayed retries used. Immediate redials are call attempts made without any delay and are considered part of a single retry. Delayed retries are made after a retry delay period.

Fax Routing. There are two generally accepted methods of placing routing information on a fax, and two kinds of routing information. Routing information can go on a separate cover page or can be added to the top of each page of a faxed document in a header band that takes up only a fraction of an inch. Routing information can consist of an alphanumeric name field or a numeric-only field.

Fax Input. How do documents get from you to your service bureau? Many schemes for this exist. You can fax a document to your service bureau using a regular fax machine. You can send documents to them on diskette, dial in to a computer bulletin board, or use special computer equipment provided by a service bureau specifically for uploading documents. Some kinds of uploading are controlled entirely by you using touch-tone entry. Other times, uploading is handled by people at the service bureau. All you need do is fax your document with instructions.

Document Creation. Can the service bureau create fax documents for on-demand transmission using digital direct techniques? Can they take files directly from your DTP applications? Nearly all bureaus should be able to do ASCII-to-fax conversion. Somewhat fewer can handle PostScript-to-fax conversion.

Database Interfacing. Many applications require interfacing to a database. Can your service bureau provide such a link? A popular database for this purpose seems to be Borland's Paradox which runs on a PC.

Fulfillment Method. You may have reasons for needing one-call or two-call fulfillment. Nearly all service bureaus should be able to do both.

Scheduling. Some applications require scheduling of delivery for a specific time of day, usually when a certain kind of information becomes available. Can requests be taken throughout the day and batched together for transmission after new information is loaded and ready?

Recurring Charges. Service bureaus typically charge either by the page, minute or tenth of a minute (6 seconds). Charges also vary depending what kind of phone line is used for the inbound call. Some typical per minute rates are shown in Table 4-2.

SERVICE TYPE	COST PER MINUTE
800 inbound calls	$0.35
900 inbound calls	$0.50
local calls	$0.16
outbound fax transmissions	$0.35

Table 4-2: Typical Service Bureau Charges

Since the recurring charges are their bread and butter, service bureaus tend to be less willing to negotiate on these charges than on the nonrecurring costs. In addition to such usage charges, there may be fixed minimums or other charges, such as a monthly fee associated with maintaining your application. Most providers offer some kind of volume usage-sensitive discount pricing, and some may have time of day variable pricing. In general, the challenge faced by a service bureau is how to maximize usage of existing facilities while avoiding overload during peak times. If you have an application that has large peaks in call volume, you may be asked to pay a capacity premium.

Non-recurring Charges. Many firms have "development fees" or "setup charges" for installing your application. These fees are usually based on the complexity of your application relative to your commitment level. A commitment level is a guaranteed minimum payment you will make to a service bureau, regardless of usage. You are in the best situation if you can point to an existing application that could be adapted to your needs with minimum modification. Costs will be highest if a lot of original work needs to be done, especially as regards programming and interfacing.

Fax-on-Demand Do's and Don'ts

- ***Avoid long complicated voice menus.*** This was covered under the section on voice menus in this chapter. It bears repeating.

- ***Structure your information to minimize depth.*** Fax-on-demand systems have one great advantage over voice-only systems: they can send long lists by fax allowing you to lay your information out in long flat lists like an index, rather than menus, submenus, and more submenus. Think catalogs and lists, not trees and hierarchies.

- ***Maximize use of fax for downstream information.*** This goes with the above point. Fax is fast, recorded voice is slow. Use print whenever possible.

- ***Tell callers what to expect.*** In catalogs or lists, give page counts on documents. If there are times when a backlog or wait for a response is possible, tell callers that as well.

- ***Use standard, not fine mode, when possible.*** Most documents, especially text, will fax well in standard versus fine mode. Some fax machines default to fine mode transmission. Override this. Use standard and save time and expense.

- *Be realistic about capacity needs.* Most people initially believe that their application will immediately generate ten-thousand calls a day or need to store libraries of documents. They buy far too much storage and line capacity than they really need. A good rule of thumb is that a single line can handle up to a hundred calls in an eight hour day, and certainly fifty comfortably. Keep in mind that in typical applications, only two percent of any given population is likely to call on any given day. Thus, a single line should be able to support a user population of approximately 2,500 potential callers comfortably for applications such as technical support or product information.

- *Know how your materials will fax.* Test everything full cycle. Fax it in and fax it out of your system. See how it looks when received. That's the only real way to tell.

- *Maintain the human touch.* Always give a phone number where a person, not a machine or recording, can be called. Give it up front and in voice as well as fax. Always remember there are people on the other end of your fax response lines.

CHAPTER 5

FAX DOCUMENT DESIGN AND PRODUCTION

DTP and Fax

Desktop publishing (DTP) is the use of computer-based tools such as Aldus PageMaker and QuarkXpress to perform the tasks of editing, layout and generation of camera-ready copy for publication. Low-cost computers, affordable software, and high-quality output devices such as laser printers have made DTP possible on even a modest budget. DTP has dramatically improved the overall quality of printed business communications.

Recent innovations in facsimile products for computers now make it possible to use these same DTP tools to create high-quality *digital direct fax* (DDF) documents. Documents sent by fax can now have an appearance approximating that of a laser printer instead of the usual low-quality fuzzy look. The connection of DTP tools and the tools used in the creation of DDF is the focus of this chapter.

The Digital Direct Fax

To understand how DTP and DDF fit together it helps to briefly review DTP, since the creation of DDF documents is just a logical extension of how DTP documents are produced.

DTP

The steps for producing documents using DTP tools can be summarized as follows:

1. Generate creative ideas for a document.

2. Create graphics and write copy.

3. Layout the copy and graphics using a DTP program.

4. Print the document on a laser printer or similar high-quality device.

With adequate design talent, the result is an attractive, clear document printed at a resolution of 300 dots per inch (dpi) or better depending on the ability of the printer. This resolution is sufficiently fine that most people, without close examination, will not perceive the dots that make up the words, lines, and objects on a printed page.

In most cases, if this document were transmitted by fax it would end up looking fuzzy and jagged. This is the familiar result of scanning by a fax machine. But by coupling DTP programs with DDF techniques, it is possible to create and transmit fax documents with quality nearly equal to that of laser-printed documents. The resolution of fax in fine mode (203x192 dpi) will produce near-laser-quality (NLQ) documents, and a well designed document transmitted in standard mode (203x96 dpi) can still have a very high-quality appearance.

DTP With DDF

A digital direct fax is created using almost the same process and tools as any other computer-based DTP document. The difference is in the last step, printing. If in the above DTP process we substitute "make or send fax" for "print," then we see where the fax part fits:

1. Generate creative ideas for a document.

2. Create graphics and write copy.

3. Layout the copy and graphics using a DTP program.

4. Use a fax utility to transmit a fax or store the fax on disk.

Instead of a printed piece of paper, the result is a fax transmission or a fax stored in a set of disk files in a raster or compressed raster format such as TIF, PCX, or DCX. If you think of a fax transmission as a remote printer, then the process is conceptually the same.

Reasons to Use Digital Direct Fax

There are several important reasons to use DDF versus simply using a fax machine to scan and transmit documents.

- *Better appearance and legibility.* This is probably the most obvious reason to use DDF versus scanning. Near laser quality faxes make a statement about your business and your communications and avoid many of the problems of fax, including fuzziness, dirt streaks, and image distortion.

- *Increased usable density.* More information can be placed on a page in the form of higher-density graphics or smaller type. Six point type is easily readable when transmitted digital direct, but when scanned it becomes so distorted that many characters are ambiguous. Newsletters, brochures, and other documents produced using DTP programs routinely employ fine type or graphics. To be useful as scanned documents they would likely need modification. Using DDF avoids the need to change.

- *Savings in transmission time.* Documents faxed DDF do not allow the introduction of "noise" due to scanning and therefore transmit faster.

Grey screens and other shaded regions will transmit faster as digital direct faxes than they would as scanned documents, although they should be avoided because they add significantly to transmission times.

- *Synergy with broadcasting and on-demand.* This is perhaps the best reason to use DDF. With some of the fax programs such as Winfax 3.0 it is far more convenient and much faster to simply "print" a copy of a document to send it as a fax than it is to take a stack of papers to a fax machine. When loading up a fax-on-demand server or broadcast system, it is also easier to copy faxes as files than to transmit documents using a fax machine. Also, the very nature of many broadcasting and on-demand applications (sales support, publishing) requires the quality that can only be achieved using DDF.

Image Quality

How much better are digital direct faxes than those sent using paper and a fax machine? The following images illustrate the image quality difference between regular fax and DDF. This first image, shown in Figure 5-1, is a test document printed on a Hewlett-Packard LaserJet 4L printer. It is text in 8-point Arial and has an effective resolution of 600 dpi. It is of the same quality one would normally expect from a laser printed document, and we will use it as our benchmark for comparison.

Figure 5-2 shows a fax of the document shown in Figure 5-1. The fax was created by taking the original and transmitting it through a fax machine in standard mode. It shows the effects of scanning and has the typical "fuzzy" fax look. Figure 5-3 shows the same document created using digital direct, again in standard mode, using Ultrascript PC from PM Ware, a special PostScript conversion utility. The original document, produced using Microsoft Word, was used to generate a PostScript file, which was then converted to a fax using Ultrascript PC. This conversion process is discussed later in this chapter.

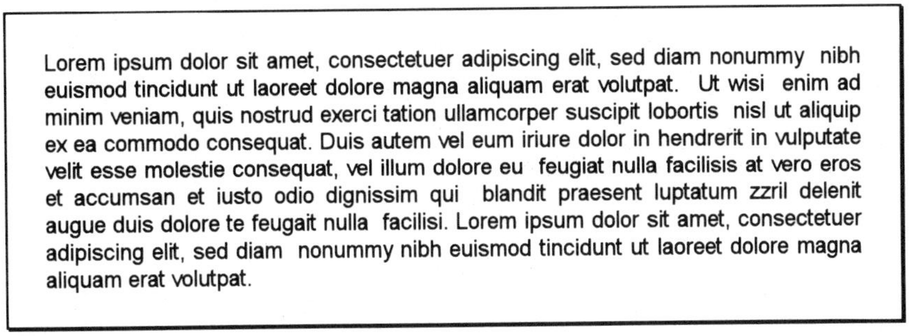

Lorem ipsum dolor sit amet, consectetuer adipiscing elit, sed diam nonummy nibh euismod tincidunt ut laoreet dolore magna aliquam erat volutpat. Ut wisi enim ad minim veniam, quis nostrud exerci tation ullamcorper suscipit lobortis nisl ut aliquip ex ea commodo consequat. Duis autem vel eum iriure dolor in hendrerit in vulputate velit esse molestie consequat, vel illum dolore eu feugiat nulla facilisis at vero eros et accumsan et iusto odio dignissim qui blandit praesent luptatum zzril delenit augue duis dolore te feugait nulla facilisi. Lorem ipsum dolor sit amet, consectetuer adipiscing elit, sed diam nonummy nibh euismod tincidunt ut laoreet dolore magna aliquam erat volutpat.

Figure 5-1: Original Test Document

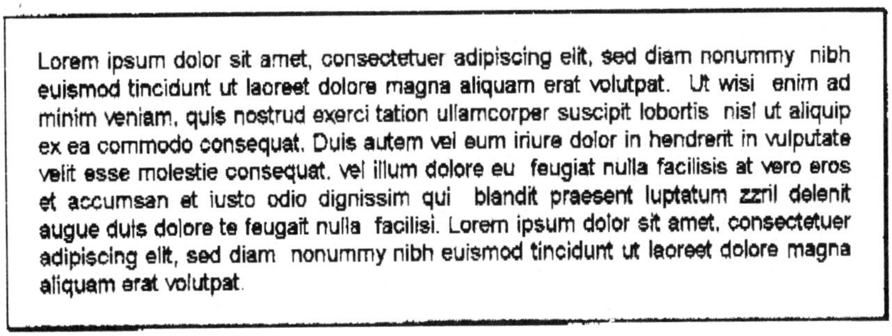

Lorem ipsum dolor sit amet, consectetuer adipiscing elit, sed diam nonummy nibh euismod tincidunt ut laoreet dolore magna aliquam erat volutpat. Ut wisi enim ad minim veniam, quis nostrud exerci tation ullamcorper suscipit lobortis nisl ut aliquip ex ea commodo consequat. Duis autem vel eum iriure dolor in hendrerit in vulputate velit esse molestie consequat, vel illum dolore eu feugiat nulla facilisis at vero eros et accumsan et iusto odio dignissim qui blandit praesent luptatum zzril delenit augue duis dolore te feugait nulla facilisi. Lorem ipsum dolor sit amet, consectetuer adipiscing elit, sed diam nonummy nibh euismod tincidunt ut laoreet dolore magna aliquam erat volutpat.

Figure 5-2: Scanned Test Document

Lorem ipsum dolor sit amet, consectetuer adipiscing elit, sed diam nonummy nibh euismod tincidunt ut laoreet dolore magna aliquam erat volutpat. Ut wisi enim ad minim veniam, quis nostrud exerci tation ullamcorper suscipit lobortis nisl ut aliquip ex ea commodo consequat. Duis autem vel eum iriure dolor in hendrerit in vulputate velit esse molestie consequat, vel illum dolore eu feugiat nulla facilisis at vero eros et accumsan et iusto odio dignissim qui blandit praesent luptatum zzril delenit augue duis dolore te feugait nulla facilisi. Lorem ipsum dolor sit amet, consectetuer adipiscing elit, sed diam nonummy nibh euismod tincidunt ut laoreet dolore magna aliquam erat volutpat.

Figure 5-3: Digital Direct Fax Test Document

Note the differences between Figures 5-2 and 5-3. The scanned image is degraded in two ways when compared to the original and the digital direct fax. It is blurred and fuzzy, and the lines that are supposed to be straight are jagged. The digital direct fax is somewhat lighter than the printed version of the test document (due to quantizing; which we will discuss later) but it is not degraded at all. The digital direct fax avoids the fuzziness and also preserves the absolute vertical and horizontal orientation of the lines in the box around the text. It does not have the jagged appearance caused by the unavoidable skew inherent in scanning. The 8-point type is clearly readable in the digital direct fax, but more difficult to read in the scanned transmission.

It is worth noting that although the resolution of fax is less than that of a laser printer, especially in standard mode, it is difficult to tell the difference between a the two when straight horizontal or vertical lines are compared, such as in the test document.

Sources of Error in Scanning

The difference in image quality between a scanned and digital fax is explained by the errors introduced into the image during the scanning process. This is what causes the fuzzy and jagged look of scanned fax. The most common errors are mechanical errors caused by dirt, clipping, and quantizing. The following sections explore each of these.

Mechanical error. The mechanical nature of the fax machine itself is a source of distortion. A fax is a collection of rollers, stepper-motors, scan heads, and other electromechanical devices, each of which introduces its own set of errors. Errors creep in because paper slips or jerks while feeding over the rollers or scan head, and because the paper is always, if only slightly, misaligned. It is effectively impossible to feed a document into a fax machine with perfect alignment. Mechanical errors include s*kew*, the rotation of a page from initial placement; *twist,* rotational movement of a page as it feeds; *slip and stretch,* the result of pinch roller slippage; and *wiggle,* which results from bending of the paper.

Dirt. Dirt can be on the paper itself or on the scan head. A piece of paper containing dirt, dust, or other particles can cause erroneous blotches or streaks to appear on the received fax. Dirt stuck to the scan head of a fax causes vertical streaks the entire length of the received fax.

Clipping. Many fax machines do not transmit portions of some pages. They usually eliminate a band at the top of the page, the bottom, or both. They may clip areas in the corners to print data such as page numbers or other fax machine information. This can result in an approximate six percent reduction in the usable length of a document. For most documents such as business letters, this is not critical, but for newsletters, product literature, or other heavily designed documents, the region of a page that is clipped can be important. Magazine articles, which have small margins, are notorious for suffering from.

Quantizing.   The greatest source of error is due to an effect called *quantizing*, or the attempt on the part of the scanning fax machine to represent what are otherwise smooth, continuous lines, letters, or other printed areas by an approximating series of black and white squares arranged on a fixed grid.

To see how this works, consider the following:

You are given a piece of ¼-inch, ruled graph paper, like the kind typically used by engineers and students.   Other than the rule lines, the paper is free from markings.   On a separate piece of paper, is a circle with a slash through it, drawn in black about the size of a dollar coin (see Figure 5-4). Your task is to take the graph paper, overlay it on the piece of paper with the circle and line, and by either leaving squares on the graph paper blank or blackening them completely, copy the circle and line onto the graph paper. Note, for each square on the graph paper, you may only totally blacken it or leave it blank, just like a fax machine would.

Figure 5-4: Now Draw This: Quantizing Test

The result would look something like Figure 5-5. As you can see, it is not possible to represent the figures accurately on the graph paper. For a curved or sloped object, you will always end up with a stairstep approximation.

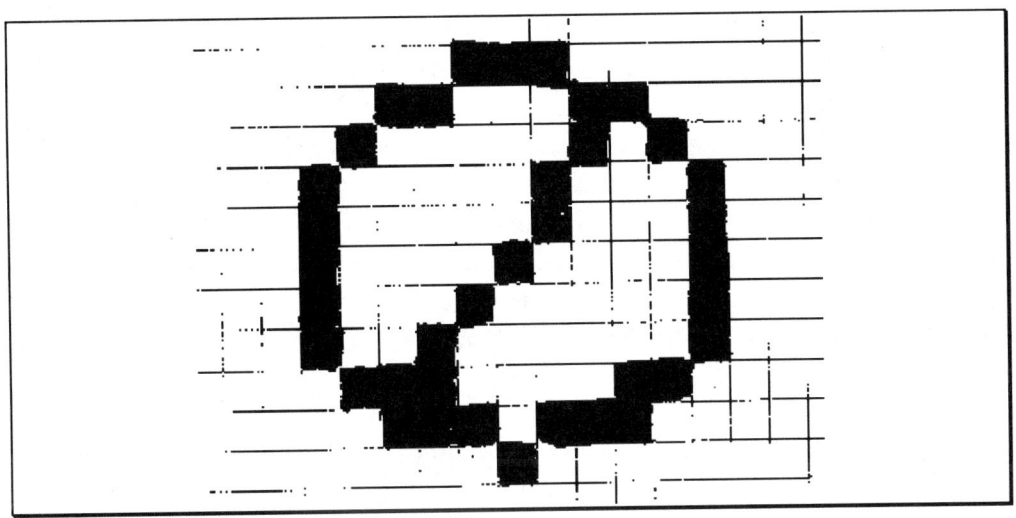

Figure 5-5: Quantizing in Action

All fax images will under some circumstances suffer from this kind of problem—but certain conditions make the results with DDF dramatically better.

With DDF, straight horizontal or vertical lines have no quantizing error except at their endpoints or uniformly along their edges. The eliminates any jagged lines or fuzziness by reducing the effect of quantizing to a level at which it cannot be perceived.

Digital direct generation of fax does not eliminate quantizing—it merely reduces our perception of it by making the errors consistent. Curved lines are still represented by a stairstep approximation. The problem of what to do with a "half pixel" still exists.   But the perception of quantizing is reduced because the stairstep pattern is consistent, not random.   "Jitter" in the image, the cause of inconsistency in quantizing, is eliminated in DDF.

Why can't the effects of quantizing be eliminated simply by feeding a document into a fax so that it goes in straight? To do this the document would need to be placed in absolutely straight and keep this exact alignment throughout the scanning process. It is virtually impossible to feed a piece of paper into a fax or scanner with such precision as to assure that lines parallel with the tops or sides of a perfectly rectangular piece of paper are in fact scanned as straight lines. It is also not possible to eliminate the errors due to the jitter from scanning. (Besides, even if you could scan with such perfection, most paper probably is not *cut* with sufficient accuracy!) The result is that there is always noticeable quantizing error.

Quantizing and lettering.   Another reason digital direct faxes look better than scanned ones is in the way they handle detailed designs and small text, which make the quantizing error more apparent. To go back to the graph paper analogy, if we try to represent a letter or small figure such as the letter "T" using graph paper, and our paper always has some skew to it, then you can imagine that the typical result will be a "T" with a stairstepped top bar and stem. Further variance is introduced by the fact that each occurrence of identical letters in a block of text will not align on the graph paper with the same relative positioning, even if all skew were eliminated. The result is that *letters are not represented consistently,* and text looks inconsistent as well as "fuzzy" when scanned.

Figure 5-6 shows a set of letters that were used in the test. These letters were then printed out on a 600dpi Hewlett-Packard LaserJet 4L and just as they are shown here. Figure 5-7 shows the result when these letters were scanned by a Panasonic PD-160E fax machine in standard mode. Figure 5-8 shows what they look like when printed as a DDF image using Winfax 3.0 and Microsoft Word for Windows 2.0a.

```
E H T L I E H T L I E H T L I
1 2 3 4 5 6 7 8 9 0 A B C D E
```

Figure 5-6: Original Test Letters

```
E H T L I E H T L I E H T
1 2 3 4 5 6 7 8 9 0 A B C
```

Figure 5-7: Close-up of Scanned Letters

```
E H T L I E H T L I E H T
1 2 3 4 5 6 7 8 9 0 A B C
```

Figure 5-8: Close-up of Digital Direct Fax Letters

Compare the close-up views of the scanned and digital direct letters in Figure 5-7 and Figure 5-8. In the scanned version, the same letters in different positions are reproduced differently. Also notice that scanning errors cause the familiar fuzziness.

With fax images generated digital direct, such problems are either reduced or eliminated. DDF eliminates angular variance; thus straight lines come out straight, and the perception of quantizing error is eliminated in letters containing straight lines. DDF also eliminates positional variance (jitter) due to scanning and thus renders each character identically.

Current fax print or conversion programs used in DDF can reproduce characters identically, regardless of placement. The variance that would be caused by how the letters are aligned relative to the scan head of a fax machine is eliminated because the fax output utility imposes a consistent representation for each character. Some earlier DDF programs, which were based on raster-to-raster translation, did not produce consistent characters, but were still better than scanned input.

Thin and Hair Lines. Rules of 3 points or less may not reproduce with consistent thickness even with DDF. This is again due to quantizing, and will be noticeable only when comparing two lines of the same width on the same document or when comparing, for example, a laser printed version with a digital fax. Use of fine mode (higher resolution) reduces this effect by a factor of two in the vertical axis, effectively decreasing the horizontal line width at which such an effect would be noticed. Vertical lines would not be affected.

Minimizing Scanning Errors. The differences between scanning and generating digital direct are minimized when:

- Most objects on the page are large.

- Nearly all lines are curved and smooth.

- There are few repeated patterns, especially small ones which are easily compared by the human eye.

An example of such an image would be a photograph, and later we will discuss how to handle them.

Digital Direct Fax Output Methods

To go from a DTP document created using a program such as PageMaker, QuarkXpress, or Microsoft Word, to a digital direct fax requires the use of either a DDF print utility or conversion utility. These utilities are separate computer programs which have the task of taking the DTP document and turning it into a fax image (DDF) which can be transmitted or stored.

There are different types of utilities that provide fax output capability, each of which has its different advantages.

Print Utilities

Print utilities are programs that allow fax transmission directly from a DTP application through a fax modem. They are designed to operate like a printer, but instead of printing on paper, they transmit the document as a fax, via a fax modem attached to the computer. Conceptually, you can think of the remote fax as the logical equivalent of a printer. To send or store a fax from a DTP document, the operator selects the fax print device and "prints" the document just as though using an ordinary printer. The transmitted document is a digital direct fax (DDF). Nearly all these utilities are designed to work with the two most common graphical user interface environments; Microsoft Windows on a PC or the Apple Macintosh operating system.

Winfax 3.0 is only one example of a fax print utility, and it is shown in the discussion that follows. There are several other print utility programs available, all of which are functionally similar. Most of these programs can use a wide variety of fax modems including internal, external, Class 1, Class 2 and CAS types. Some come bundled with fax modems, while others are available separately. The choice of modem is almost completely independent of the software. Sources for these programs are listed in the Appendix.

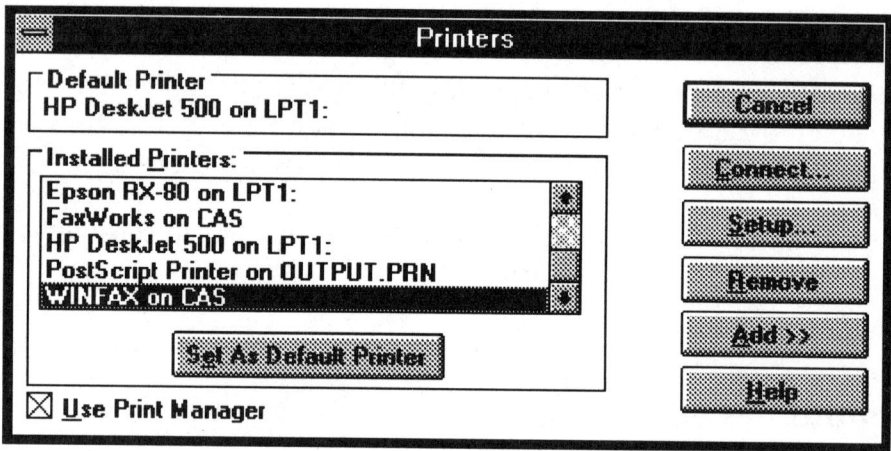

Figure 5-9: Selecting a Fax "Printer" in Windows 3.1

To transmit a fax instead of print a document on a printer, the fax device must first be selected as the target printer. In the Windows 3.1 environment, the fax "printer" is selected just like any other printer would be, using the Control Panel Printer or the "Print setup..." part of an application. The Control Panel Printer window shown in Figure 5-9 gives several choices; the fax, "WINFAX on CAS" has been selected as the target printer.

Once "WINFAX on CAS" is selected, a print command brings out the normal print window as shown in Figure 5-10. Clicking on the OK button then brings out the Winfax 3.0 send window as shown in Figure 5-11. At this point the fax print utility program Winfax 3.0 has taken over. If we had selected "FaxWorks on CAS" then we would instead see the FaxWorks 2.50C Select Destination window as shown in Figure 5-12. At this point, these utilities allow the operator to select or enter a fax number, recipient, cover sheet text, time of transmission, and resolution, as in the example from Winfax 3.0. They duplicate the essential functions of a fax machine, except without the paper!

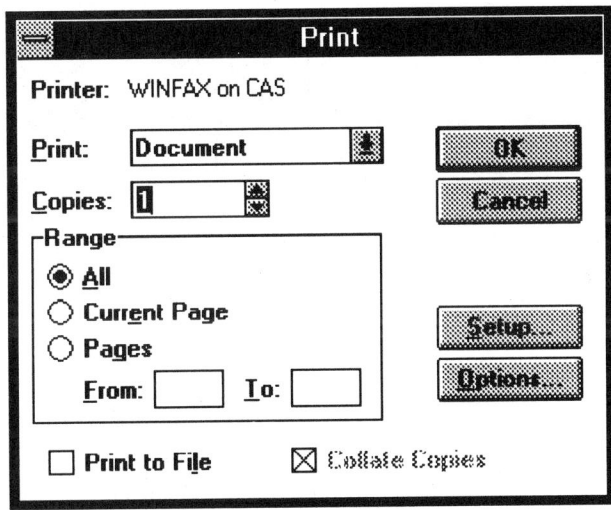

Figure 5-10: Normal Windows 3.1 Print Window

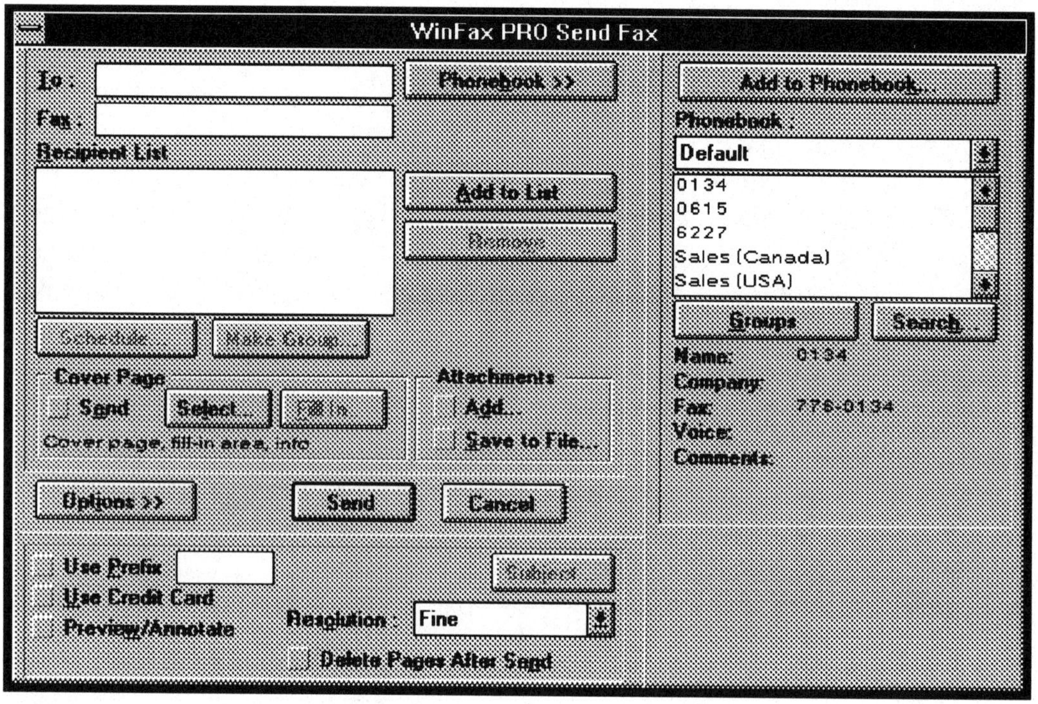

Figure 5-11: Winfax 3.0 Fax Send Window

Print utilities have many features to make routine tasks faster. For example, entering a phone (fax) number is required in order to transmit a document. As a convenience in entering frequently used numbers, most utilities have a "phone book" feature which stores these numbers in a directory for repeat use. Most phone books require the user to manually enter the names, numbers and other data from the keyboard. But because these numbers are likely to already be in a database, Winfax 3.0 goes one step further. It allows loading its phone book from a dBase III/IV file, a CAS file, a phone book file from the previous version of Winfax 2.0, or a delimited ASCII format file. This feature makes it easy to transfer fax numbers from a database into the Winfax 3.0 phone book. For example, it is easy to quickly transfer several hundred names and fax numbers from a Paradox 3.5

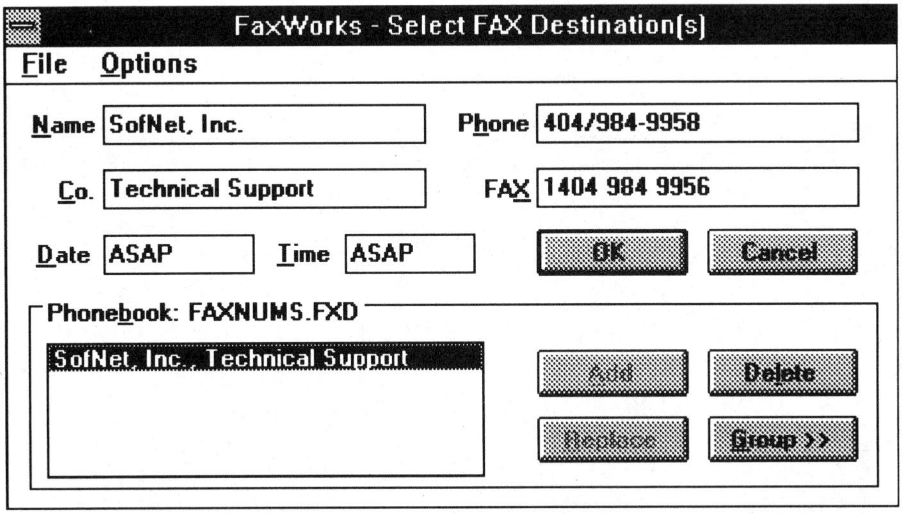

Figure 5-12: FaxWorks 2.50C Select Fax Destination Screen

database to Winfax 3.0. Figure 5-13 shows the Winfax 3.0 fax number import screen.

From the Winfax 3.0 Send Fax window shown in Figure 5-11, selecting recipients is simple. The user double-clicks on a phone book entry (such as Sales USA) or drags the entry to the Recipient List box (on the left) or clicks once on the entry and presses the Add to List button. For large lists, the search feature allows selection of phone book entries by name, company, or any of the fields it holds. This can be much faster than scrolling manually through the list.

Fax broadcasting is easy using Winfax 3.0. For transmissions to a group of people, broadcast lists can be created using the Group feature. The program will then automatically select all the numbers in the group when the group is selected. It will then control the fax modem to sequentially dial and transmit the document to each of the recipients in the group.

Figure 5-13: Winfax 3.0 Fax Number Import Feature

Transmission is started by clicking on "Send" in the Fax Send menu starts the transmission. With Winfax 3.0, transmission begins after a few seconds, and can work in the background. Transmission status is displayed as shown in Figure 5-14 and can be hidden by minimizing the Winfax 3.0 window. Operation with a CAS-compatible modem such as the Intel SatisFAXtion 200 is smooth.

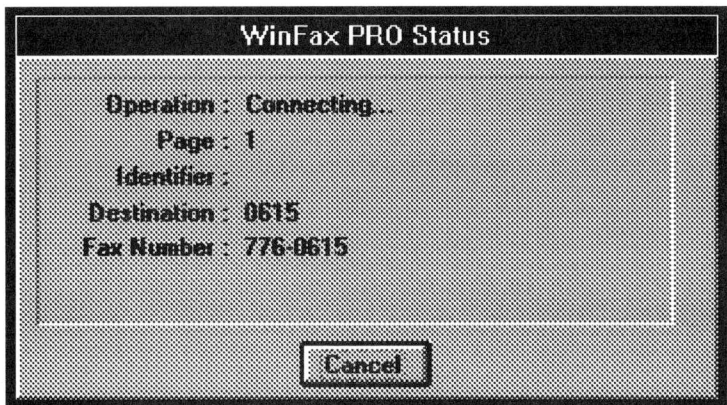

Figure 5-14: Winfax 3.0 Transmission Status

Printing to a file. Printing to a file is how a digital direct fax of a document is created and stored on disk for later transmission, viewing, or printing. For documents that may be created or revised infrequently, this can a very convenient way to save time. The convenience and time savings are possible because once a document is stored as a DDF, it is no longer necessary to open the original document and repeat the print to fax steps outlined earlier in order to send it by fax. Instead, it is only necessary to use the fax print utility, Winfax 3.0 in this case, to direct the transmission of the already stored fax. Since the original document and its associated application are not involved in the fax transmission process, there is an additional benefit to using stored DDF documents—they can be sent by anyone with only the matching fax print utility. The sender would not need the original document or its application.

Consider the difference between the following two sequences:

Standard Print to Fax Steps:

1. Open the original document using its associated application.

2. Select the fax as the print device.

3. Print the document to the fax.

4. The fax transmission application opens.

5. Enter the fax transmission information and start transmission.

Sending a Stored Fax:

1. Open the fax transmission application.

2. Select the stored fax to transmit.

3. Enter the fax transmission information and start transmission.

Winfax 3.0 handles this print to file task with a check box in the Fax Send window (Figure 5-11) and a Save to File subwindow, shown in Figure 5-15. To print to a file instead of send a transmission, you check the Save to File box in the Attachments area of the Fax Send window and press the Send button. This brings up the Save to File subwindow (see Figure 5-15), which allows you to enter descriptive information that is useful for finding and selecting stored documents. Stored documents are then sent by attaching them to a transmission, which could consist of nothing more than a cover page. Attachments are stored in specially indexed files and are managed using an attachment management facility in Winfax 3.0.

Export. Nearly all print utilities such as Winfax 3.0 use an application-specific file format for faxes that are sent, received, or stored. These internal file formats are not standard and cannot be directly read by most other programs. Exporting from a print utility is one way to convert these faxes into other graphics file formats that can be used by other programs such as draw or paint applications common to DTP. Common file formats are ones such as PCX, DCX, TIF, or BMP. Another way to convert fax files to other graphics file formats involves using a specialized raster conversion program such as Hijaak, discussed later in this chapter.

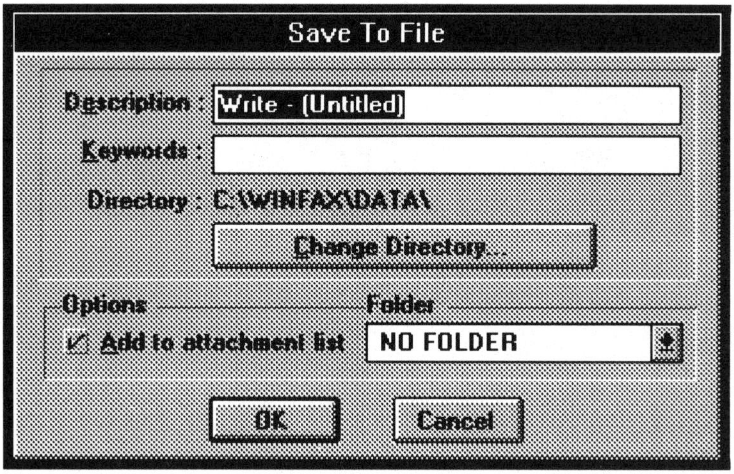

Figure 5-15 Saving a Fax Image to Disk Using Winfax 3.0

Winfax 3.0 exports faxes from the same window used to view documents. (See Figure 5-16) This window is used when viewing sent, received or stored fax documents. Under the File menu of the viewer is a selection for Export. Selecting Export brings up the Export window, shown in Figure 5-17. In this window, under File Format, there are three choices for standard file formats.

Fax print utilities are the simplest way to send and receive faxes as well as create and store DDF documents. To go from document to fax, they work just like a printer and are therefore compatible with any application that has a print feature. To go from fax to DTP program, print utilities provide an export feature that will convert a fax document into a file in one of the more common graphics formats. They offer high-quality when operating in fine mode, but do not perform quite as well as fax conversion utilities when operating in standard mode. For more information on print utilities, see the resource directory in the Appendix.

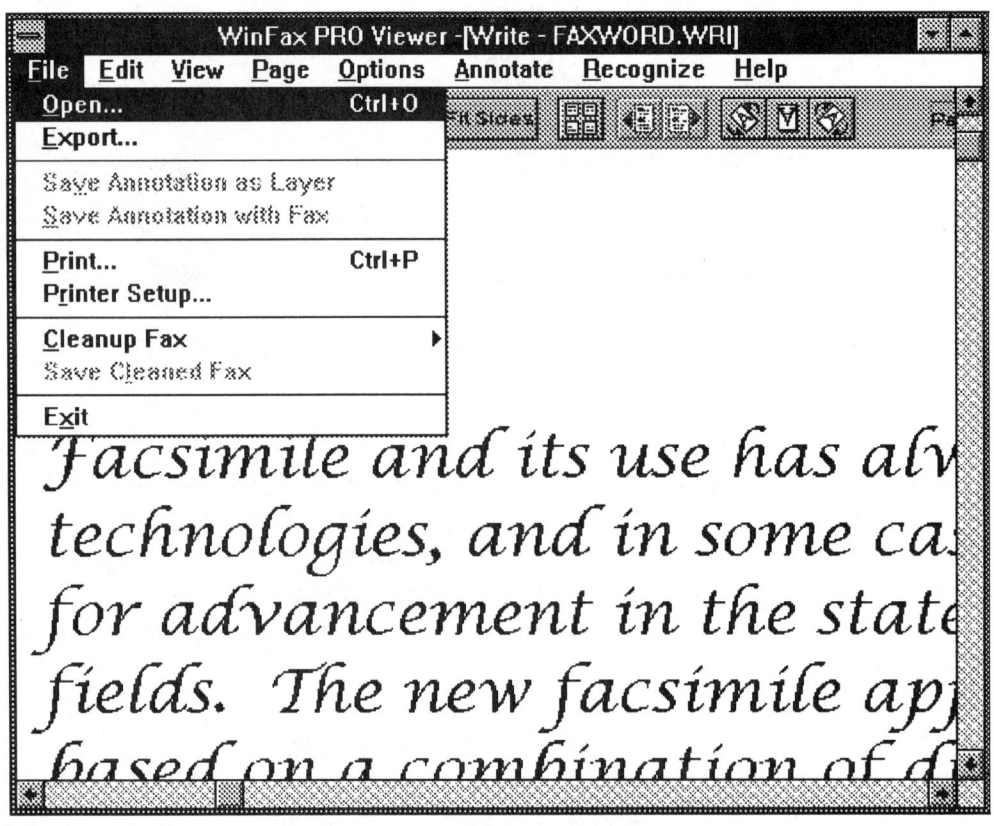

Figure 5-16: Winfax 3.0 Fax Viewer

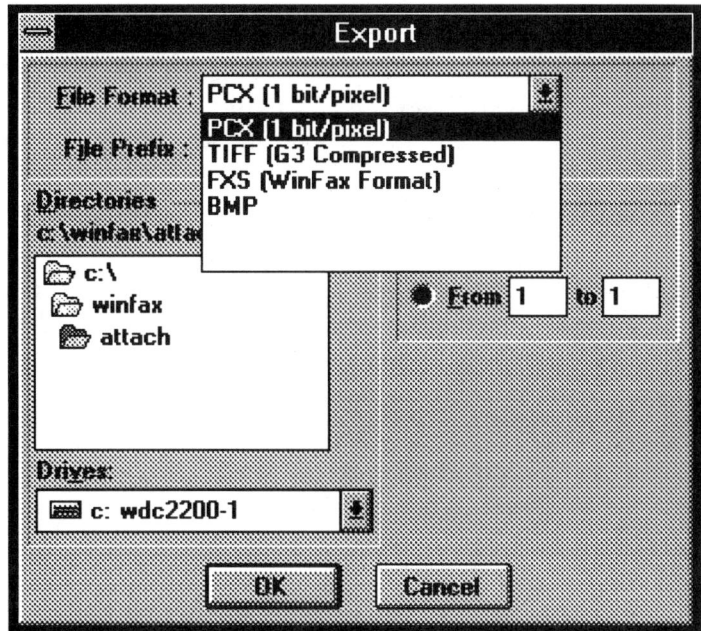

Figure 5-17: Winfax 3.0 Fax Export

Fax Conversion Utilities

Before fax print utilities for graphical environments like Windows were available, conversion programs were the only way to take documents created using DTP programs and turn them into digital direct faxes. While performing conversion through some external means may seem unnecessary or even undesirable given the availability of print utilities that run in environments like Windows, there are still some important reasons one might want to use a conversion rather than a print utility.

The most significant reason to use a print utility such as Ultrascript PC instead of a print utility such as Winfax 3.0 is to get the best possible fax document quality when operating at the limit of resolution and feature sizes. For example, Ultrascript PC significantly outperforms the print utilities when creating fax documents in standard mode. A standard mode fax of a document using 12 point Times New Roman looks significantly better when produced by a conversion utility such as Ultrascript PC than when produced by Winfax. The difference is much less noticeable when using fine mode.

An important secondary reason to use a conversion utility is to preserve the layout of a document that will be both printed and faxed as a DDF. Conversion utilities tend to generate the same spacing and layout as most printers. Print utilities, however, tend to cause line and page breaks to fall in different places when a document originally composed for a printer is retargeted for fax via a print utility. These effects are discussed later on in this chapter.

Another reason to consider using a conversion utility is to accommodate the need to operate in a cross-platform, cross-environment, or multi-application environment where print utilities are unavailable or incompatible with the applications used in document creation. We will cover this in greater depth later in this chapter.

Say you want to produce a small print document like *BioWorld Today*. There are two factors that will require a tradeoff. The first is print quality and the other is transmission time. *BioWorld Today* is produced in standard resolution and is operating near the limit of performance for standard mode and the type size used. Both quality and transmission time are important here—quality is important because readers want a publication that is easily readable, and time because it directly affects one of the largest costs of publishing the newsletter.

BioWorld Today found that the best solution was to use Ultrascript PC, a conversion utility. This program provided the maximum type density, at the best quality, when creating a standard mode DDF. It was used to convert a PostScript file from QuarkXpress to a standard resolution DDF. Experiments were done using an Abaton InterFax fax modem and its associated print utility software, and the type quality in standard mode was not nearly as good as with Ultrascript PC.

We can see an example of this if we take a test document in Times New Roman 8-point type and generate standard resolution DDF documents using Ultrascript PC and Winfax 3.0 and compare them. Figures 5-18 and 5-19 show the test document as produced by Ultrascript PC and Winfax 3.0, respectively. It is obvious that the former is legible and the latter is not. Of course, in a real publication you would not want to push the limits this far, but the examples do illustrate the performance differences.

Further experimentation with different documents and text styles and sizes would show that as type size and graphic feature size increases, the differences in performance between print and conversion utilities diminish. Beyond 14-point type the differences are not very noticeable even in standard mode. In fine mode, the differences are still noticeable, but probably not beyond 10-point type, and that is probably as small as you would ever want to fax.

Conversion programs work by taking one of the following three kinds of inputs and converting them to fax format files:

- PostScript

- Raster files (e.g. PCX)

- Printer data streams (e.g. PCL)

Programs that convert PostScript to fax, such as Ultrascript PC from QMS, Inc. (marketed by PM Ware) and GoScript from LaserGo, Inc., were originally designed to act as host-based PostScript interpreters to drive non-PostScript printers. Because the process of converting PostScript for non-PostScript printers is essentially the same as converting PostScript into a variety of raster formats, adding fax as another type of output was a relatively simple and logical extension of conversion for printers.

Lorem ipsum dolor sit amet, consectetuer adipiscing elit, sed diam nonummy nibh euismod tincidunt ut laoreet dolore magna aliquam erat volutpat. Ut wisi enim ad minim veniam, quis nostrud exerci tation ullamcorper suscipit lobortis nisl ut aliquip ex ea commodo consequat. Duis autem vel eum iriure dolor in hendrerit in vulputate velit esse molestie consequat, vel illum dolore eu feugiat nulla facilisis at vero eros et accumsan et iusto odio dignissim qui blandit praesent luptatum zzril delenit augue duis dolore te feugait nulla facilisi. Lorem ipsum dolor sit amet, consectetuer adipiscing elit, sed diam nonummy nibh euismod tincidunt ut laoreet dolore magna aliquam erat volutpat.

Figure 5-18: Ultrascript PC at the Limit

Lorem ipsum dolor sit amet, consectetuer adipiscing elit, sed diam nonummy nibh euismod tincidunt ut laoreet dolore magna aliquam erat volutpat. Ut wisi enim ad minim veniam, quis nostrud exerci tation ullamcorper suscipit lobortis nisl ut aliquip ex ea commodo consequat. Duis autem vel eum iriure dolor in hendrerit in vulputate velit esse molestie consequat, vel illum dolore eu feugiat nulla facilisis at vero eros et accumsan et iusto odio dignissim qui blandit praesent luptatum zzril delenit augue duis dolore te feugait nulla facilisi. Lorem ipsum dolor sit amet, consectetuer adipiscing elit, sed diam nonummy nibh euismod tincidunt ut laoreet dolore magna aliquam e ... volutpat.

Figure 5-19: Winfax 3.0 at the Limit

Utilities like Hijaak from Inset Systems take raster files or printer data streams as input perform a raster-to-raster conversion to generate fax output. For example, they can take output intended for an HP LaserJet printer and convert that to a fax or other raster format such as PCX. Similarly, they might take a file produced by a paint program in PCX format and convert it to a fax. The basic concept is that they are running conversions between two different raster formats.

Hijaak is an interesting program because it is apparently the only utility intended to provide conversions between a wide range of graphic and fax file formats. Hijaak is available for both the DOS and Windows environments. In the Windows environment, Hijaak allows viewing, capturing, converting, processing, and printing of both vector and raster images. It supports the Windows Clipboard, object linking and embedding (OLE), drag and drop, and dynamic data exchange (DDE), and TrueType fonts. It provides conversion between more than sixty formats including fifteen vector, twenty-three raster, and twenty-four fax formats. Both versions of Hijaak are highly recommended for anyone who expects to do extensive work with graphic format files. These tools can sometimes be just the trick that is needed to make an unusual conversion that might otherwise be impossible or extremely difficult.

Most conversion programs seem take one of two approaches to obtaining input. The first strategy is to a process a file that was created by another program. This could be PostScript that was placed in a file, or a printer data stream that went into a file rather than to a printer. The second approach is to capture the data stream intended for the printer port of the computer on-the-fly and perform a conversion without any intermediate file. In the latter case, the output from an application such as a word processor or spreadsheet program would be diverted to the conversion utility instead of to the printer. Typically these programs use a TSR (terminate-and-stay-resident) program that intercepts the data stream normally intended for a printer. When the data stream is intercepted, the TSR, perhaps in conjunction with another program or set of programs, takes printer output and converts it to one or more different formats, including fax image files.

Ultrascript PC from QMS, Inc. Ultrascript PC is a program that both captures and processes approaches. It is one of the best programs for converting PostScript (EPS) files to practically any printer or raster format, including fax. QMS, Inc. also manufactures a line of high-performance PostScript laser printers, which may be part of the reason why Ultrascript PC renders fonts with consistently high-quality. Version 3.01 is fully compatible with TrueType fonts under Windows 3.1.

Ultrascript PC is marketed as a utility that lets non-PostScript printers print PostScript documents by using the processor in the PC to perform the conversion between PostScript and whatever raster format is required by a wide variety of dot matrix and similar printers. Most users of Ultrascript PC will use it in this manner. What is undocumented and unfortunately not promoted as one of the benefits of the program is that with some tinkering it is very capable of turning EPS files into very high-quality fax TIF files that are directly compatible with Gammalink's Gammafax fax modems, and into PCX and DCX files that can be used by other fax modems such as the Intel SatisFAXtion. In what follows we will cover just how to get Ultrascript PC to produce fax files.

When you first install Ultrascript PC and begin using it, you must identify the type of printer you have connected to your computer. The screen from which this choice is made is located under File, Printer Setup and is shown in Figure 5-20.

Missing initially from the list of printers and other output devices are DCX and TIF format files. A selection for PCX format files and fax resolution should exist. But before fax TIF and DCX formats are available, a special modification to one of the control files which needs to be made before these selections will appear. These features are already part of the program, but they're not documented and are shipped disabled.

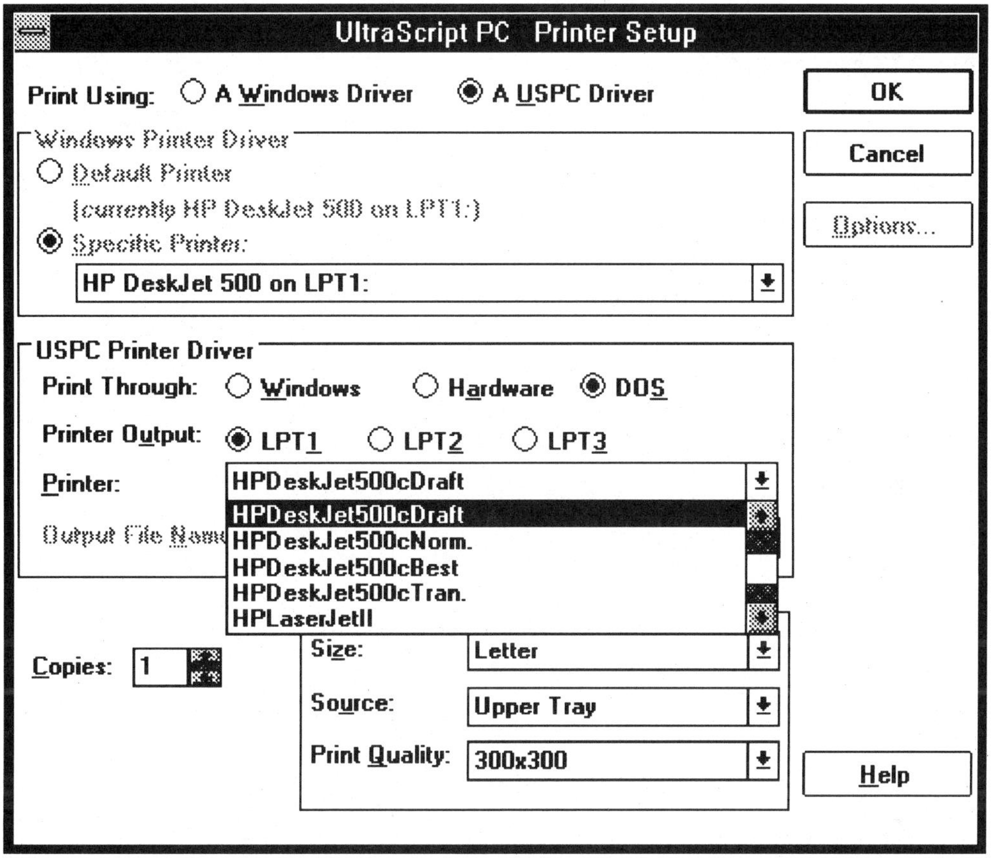

Figure 5-20: Ultrascript PC Printer Selection Window

Modifying Ultrascript PC. Here is how to modify Ultrascript PC to get TIF files and DCX files that can be used as fax images with products such as the Gammafax CP and Intel SatisFAXtion.

Ultrascript PC is usually installed in a directory named \USPC. Go into this directory and find a file named `drivers.cfg`. Open that file with an editor able to handle a file of that size, such as the Norton Editor or the editor that comes with DOS 5.0. Search for a set of lines that look like those shown in Figure 5-21.

```
# Compressed TIFF format files (Type 3)
# (Only useful for FAX machines)
#>TIFF_fax:203x192:8500x11000:FF:(0,0)(0,0)
#>TIFF_fax:203x192:8500x14000:FF:(0,0)(0,0)
#>TIFF_fax:203x192:8270x11690:FF:(0,0)(0,0)
```

Figure 5-21: Original Ultrascript PC DRIVERS.CFG

Replace these lines with the lines as shown in Figure 5-22.

```
# Compressed TIFF format files (Type 3)
# (Only useful for FAX machines)
>TIFF_fax_std_ltr:203x96:8500x11000:FF:(0,0)(0,0)
>TIFF_fax_fine_ltr:203x192:8500x11000:FF:(0,0)(0,0)
>TIFF_fax_fine_lgl:203x192:8500x14000:FF:(0,0)(0,0)
>TIFF_fax_fine_A4:203x192:8270x11690:FF:(0,0)(0,0)
```

Figure 5-22: Modified Ultrascript PC DRIVERS.CFG

Save the file to disk. Now reboot your machine and restart Ultrascript PC and go back through the "Printer Setup" menu to the "Printer Selection" window you should be able to pick one of the above print devices. Your selection list should look like the one shown in Figure 5-23.

To add the ability to generate DCX files, look for the lines in the drivers.cfg file that look like the ones shown in Figure 5-24, and remove the leading # character from each line, then save the file.

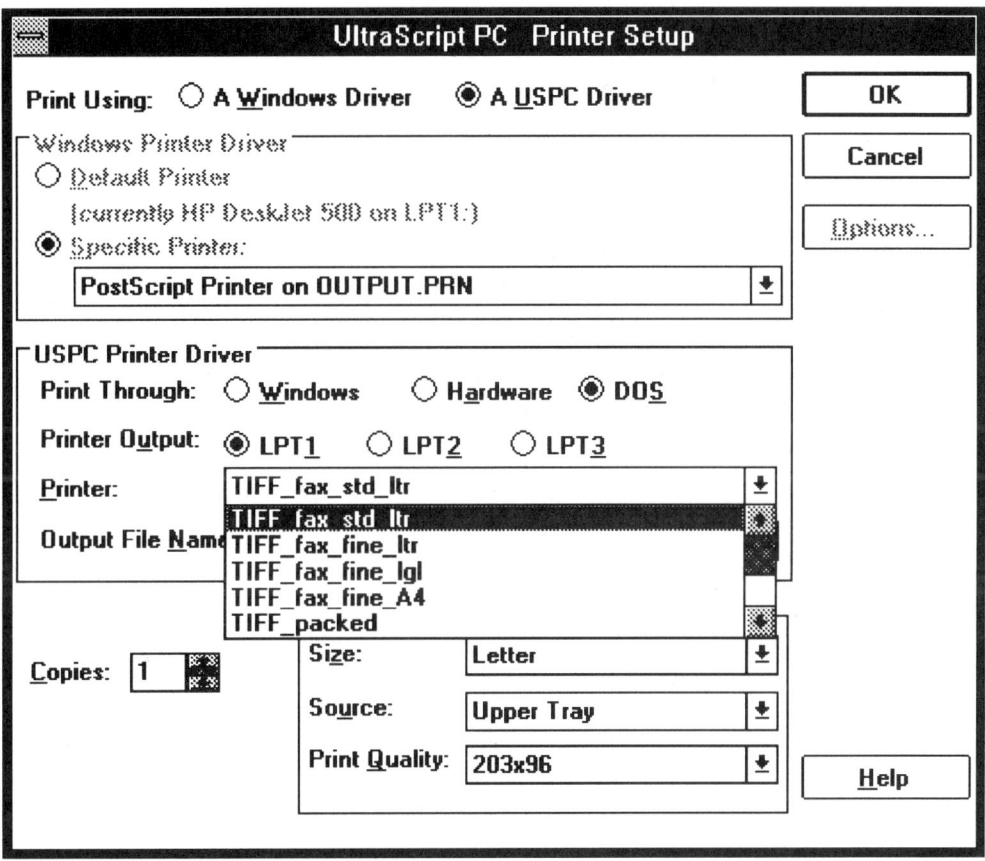

Figure 5-23: Revised Ultrascript Printer Selection Window

```
# DCX format
#>DCX:203x192:8500x11000:WF:(0,0)(0,0)
#>DCX:203x192:8500x14000:WF:(0,0)(0,0)
#>DCX:203x192:8270x11690:WF:(0,0)(0,0)
```

Figure 5-24: DRIVERS.CFG DCX File Format Entry

Your copy of Ultrascript PC should now convert EPS files to TIF, PCX, and DCX format compatible with Gammafax CP, Intel SatisFAXtion, and other fax modems and DTP programs. Note, when printing to PCX or DCX files, be sure to set the Print Quality to 203x192 (fine) if you intend to transmit this file using a fax modem. Fax modems generally expect DCX and PCX format files to be in fine resolution. Sending a PCX or DCX file with standard resolution will result in a file that appears shrunken by 50 percent in the vertical direction. Use of any other resolution may result in stretched, compressed, garbled, or clipped images. Your choices for page sizes and resolutions are letter size, fine and standard, and legal and A4 in fine only. If you want legal and A4 in standard, add two more lines to drivers.cfg as shown in Figure 5-25.

```
>TIFF_fax_std_lgl:203x96:8500x14000:FF:(0,0)(0,0)
>TIFF_fax_std_A4:203x96:8270x11690:FF:(0,0)(0,0)
```

Figure 5-25: DRIVERS.CFG Modified for Legal and A4 Paper Sizes

To use Ultrascript PC to create fax TIF or PCX file:

1. Select the appropriate "printer" using Print Setup.

2. Use the Print window to tell Ultrascript PC to begin converting the EPS file to a TIF file.

3. When completed, the TIF files are given names such as US001.TIF, US002.TIF, and so forth. They are typically placed in a directory named \USPC\fileoutp.

Once these fax images are on disk, you can view them using a viewing utility—such as FAXD.EXE which comes with the Gammalink fax boards—or you can simply send them to a fax machine and look at the result. The latter is the ultimate proof of what the end result will look like. Using viewing utilities is convenient, but because of the difference between screen resolution and fax resolution, what you see on a screen will not be exactly what you see on paper. Ultrascript PC does not provide any way to view a fax TIF file.

There is a bug in Ultrascript PC Version 3.01 that affects the creation of fax images as described in this example. The program will not print any pages beyond the first one in a document when run under Windows. It will create zero-length files for any pages after the first. There are two ways to work around this.

The first is to print a document to separate EPS files, one page at a time. This can be tedious, but if the document is short and you're proofing one page at a time anyway, it's not a bad way to go. The other method is to print a document to an EPS file, exit Windows, and start up Ultrascript PC under DOS. This is much better if you are printing a document with quite a few pages. You avoid handling separate files for each page, and Ultrascript PC is very processor-intensive and will run a good bit faster when running only under DOS. To run Ultrascript PC under DOS, you go to the \USPC directory and type USPC.

Ultrascript PC, like other programs that process EPS data, is processor-intensive. Performance is unacceptably slow on 286 and slower 386 machines. A faster 386, such as a 33 Mhz machine, should be considered the minimum acceptable performance level. Given current technology costs, this is not asking much. Adding a 387 math coprocessor or using a similar speed 486 machine should give about another 30 percent performance improvement. This is one application where having a faster computer makes a big difference. Table 5-1 shows some performance data to give you an idea of how long conversions take. The times shown indicate the time to convert an EPS file of text, similar to a page in this book, into a TIF file.

The test was performed on a 386/40 Mhz AT-compatible clone PC with 8Mb of memory that runs the Landmark performance test at 61.33 and the QA Plus system performance test at 11379 Dhrystones and 140.9 Whetstones. Ultrascript PC was initialized before conversion to remove the delay associated with program start-up. In all cases the same `config.sys` file and `autoexec.bat` file were used. By comparing the performance numbers of this test machine with those of a machine you might use for this task, you can get some idea of expected performance.

ENVIRONMENT	DOCUMENT SIZE	RESOLUTION	SECONDS
Windows 3.1	U.S. letter	standard	20
Windows 3.1	U.S. letter	fine	35
DOS 5.0	U.S. letter	standard	14
DOS 5.0	U.S. letter	fine	20

Table 5-1: Ultrascript PC EPS to Fax TIF Performance

Generating PostScript Files from Windows. The easiest way to generate PostScript output from a Windows application is to create a PostScript print device and then print to it. Setting up Windows to generate PostScript files is straightforward, but not well documented. Use the following steps.

1. Open the Control Panel, then open the Printers item.

2. Install a PostScript printer by clicking on the ADD button, highlighting the PostScript Printer line in the List of Printers, and clicking on the Install button.

3. The PostScript printer should now appear on the list of Installed Printers.

4. "Connect" the PostScript printer to the FILE output device by pressing the Connect button and highlighting the FILE: port from the list of ports.

5. Press the OK button and the printer is installed.

6. Before using the printer you must specify the name and location of the PostScript output file.

7. Click on the Setup button to get a screen that says "Apple LaserWriter Plus on FILE:"

8. Click on Options and click on the switch for Encapsulated PostScript File.

9. Fill in the full path and file name of the file that you want to contain your PostScript output. One suggestion is to use a temporary file name that does not change, and rename the file, since that is less cumbersome than constantly changing the setup. (For example: \tmp\output.ps.)

GoScript. GoScript from LaserGo, Inc. is a PostScript language interpreter for the PC. Like Ultrascript PC, it is intended to take PostScript language text and graphics and print them on laser, ink jet, and dot matrix printers.

GoScript comes preconfigured to support some fax format files. Among those files are:

- Fujitsu dexNET PC210 DXN format

- Touchbase WorldPort 2496 WFX format

- Intel SatisFAXtion DCX format

- PCX format in fax resolutions

- TIFF format in fax resolutions (not Gammafax-compatible)

- PCL format in fax resolution for use with Hijaak in converting to other fax file formats.

GoScript will produce letter, legal, and A4 sizes in fine and standard resolutions. The current version comes with thirty-four outline fonts that closely correspond to fonts available from the Apple LaserWriter PostScript font set. Additionally, GoScript is compatible with Adobe Type 1 and 3 PostScript format fonts.

The Multifunction Computer-Fax Machine

An important type of fax machine is just starting to attract significant market attention. Available since 1985 but not in wide use, such machines are just now starting to enjoy wider distribution. This new type of fax machine is often called a *multifunction fax machine* (MFM) because it is able to act as a scanner, printer and fax all at once. These machines connect with computers and work with special computer software. This combination makes them very useful as desktop publishing tools. Additionally, the computer software that drives the machine makes it easier for the user to control and benefit from the advanced features of these sophisticated fax machines.

The key to these machines lies in their connection with a computer that runs special MFM software. Typically, an MFM is connected to a computer using a commonly available RS-232 serial interface cable. Some very advanced machines, such as the Canon FAX-L6500, which sells for $29,900, have higher-speed interfaces such as SCSI, video, or a proprietary method. This interface allows the MFM software to control the fax machine and make use of its scanner, printer, fax and other capabilities such as on-board memory.

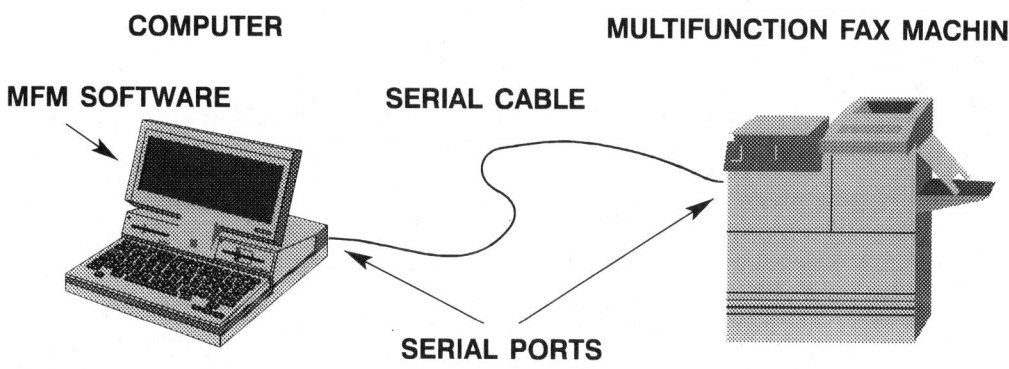

Figure 5-26: The Multi-Function Machine and Computer Connection

Several MFM machines are already in the market and more are planned for introduction. This is sure to be a hot area for product development, since it is at the confluence of two large industries: office equipment (fax) and computers. Canon produces a line of serial port fax machines including the latest and most powerful L775 laser engine fax, the L770, T-701, T-301, and FAX-210. As one of the early providers of MFM products, Canon enjoys the wide support of several MFM software packages. Toshiba offers the TF-511, Relisys has the TeleFAX RA-2136, and Okidata has just introduced the new DOC*IT 4000 PS. Murata has recently introduced the

F-75, and NEC has been selling its printer-turned-fax as the Model 97 and 95.

Each of these MFM have different strengths based upon very different features. There does not seem to be one "best" machine for any given task, as each offer a mix of benefits. Many of the machines come with their own computer software, some require software from an independent vendor (Canon, for instance, does not offer its own software), and others offer both options. They all work with a computer to provide the multiple benefits of faxing, scanning, and printing, and can be very useful in the production of digital direct fax documents. Figure 5-26 illustrates the MFM-computer connection.

To give you some idea of what sophistication can be packed into an MFM consider the following, which is a summary of the Canon L770 features:

- *Fax.* Documents created on the computer can be sent over a directly connected cable from the computer to the fax machine for transmission. There is no need to use regular transmission to transfer the document from computer to fax. This is a digital direct fax. Documents received by the fax can be sent from the fax directly to the computer without the need to print them first. This allows direct capture of a fax by the computer, similar to using a fax modem. Received documents may be printed or just stored in the fax or computer. All fax functions can be controlled by the computer and the fax can operate as a normal, unattached unit.

- *Printer.* Documents received by the fax, scanned into the fax, or created on the computer can be printed by the fax, similar to how a computer printer would perform the same task.

- *Scanner.* The scanner on the fax scans sixty-four levels of gray in letter, legal, and other sizes. It scans faster than fax transmission, at a rate of five seconds per page in standard, fine, and super-fine resolution, regardless of content. (The higher performance machines, such as the

Canon L785 and L775 will scan at two seconds per page.) A special UHQ chip improves representation of gray levels with a proprietary Error Diffusion Method. This method places dots randomly, rather than systematically, when simulating gray areas on a page. The result is a high-quality continuous appearance to gray tones, versus the usual lumpiness typical of dithered patterns. UHQ also provides automatic intensity control. Documents scanned by the fax can be transmitted, printed, or transferred to the computer.

- *Automatic Document Feeder.* The fax has a document feeder with a thirty sheet capacity. Coupled with the memory, this feature makes feeding in multipage documents very fast and convenient. Hand-feeding single sheets into a slow scanner is time consuming and inadequate for any but the lowest volume applications.

- *Memory.* The fax can store documents that were received or scanned or transferred from the computer. A machine may have from 768k to 3.75M of total capacity for receiving or storing fax documents in up to ninety-nine discrete locations called *electronic file folders (EFF)*. With memory, any document that passes through the fax can be held in memory so that it can be downloaded, printed, or transmitted under computer control.

- **Broadcasting.** The fax will accept lists of numbers from the computer for broadcast operation. With software such as CAN-FAX from L.A. Business Systems, each document can be "mail merged" to look like it was typed and signed individually by the sender.

Multifunction Fax Machines versus Computer Peripherals

There are many reasons to use an MFM versus a separate scanner, printer, and fax modem, some of which are related to advantages in the creation of digital direct fax documents and others are based on cost or ease of use.

Cost. The cost of a Murata F-75 is approximately $1,500 in the U.S. right now. It prints on plain paper; scans in standard, fine, and superfine resolutions; will transmit at 9,600 bps with ECM to Group 3 and to Group 2 machines; comes standard with 512k memory; and has a thirty page document feeder. It also has a built-in line sharing device for voice and fax line sharing. It will work either with its own PCAS software or with SciFAX F-75 PC which retails for $199. Taken separately, the cost of a printer, scanner, and fax modem would be about $3,500. For even greater economy in an MFM, the Toshiba TF-511 is only about $1,100.

Space.   For small businesses and home offices, space can be a consideration. A laser printer and scanner require about twice the space as a fax machine.

Imaging Quality.   Two factors affect image quality when comparing scanners with MFM.   One is the scanned resolution and the other is the handling of halftones. Fax machines scan at fax resolution and there is no conversion. Scanners typically scan at resolutions of 300 or 600 dots per inch. In converting between scanner and fax resolution, aliasing error is introduced, increasing the "fuzziness" introduced by scanning. Many MFM have built-in processing to handle the special job of representing gray-scale images as binary images in a fax.   The Canon L775, with its UHQ-II circuitry, produces excellent fax images of photos. Using a separate scanner would require software of comparable quality on the computer to perform the same job.

Proofing. What you print with an MFM is what you fax. Proofing with a fax modem requires transmitting the document from the fax modem to a fax machine. Printing on an MFM is proofing.

Paper Handling. Many documents are still created by the time-honored cut and paste (on paper versus on computer) method. For a variety of reasons, this is often the fastest or best accepted way of creating documents. Assuming such documents are to be turned into a fax for transmission, rather than created originally on a computer, an MFM can be much faster than using separate computer peripherals or a noncomputer-controlled standalone fax. It can scan in such a document, store it in memory, and immediately begin transmission much faster than if a separate scanner and fax modem had been used. CrimeFAX, discussed in Chapter 2 is an example of this.

Error Correction. Most MFM have ECM. This feature, described in Chapter 2, prevents problems such as scan line drop out from happening during transmission. As of this writing, only one fax modem, the Gammafax CP, is known to have ECM.

Transmission Performance. Many MFM such as the Canon machines, will transmit using MMR compression, resulting in a thirty percent to fifty percent reduction in transmission time. Some machines may also have higher transmission speeds including 14.4k bps. Few fax modems have higher transmission speeds, and none are known to have MMR compression.

Interoperability with Proprietary Features. An MFM is fax machine, not just a modem, and depending on the features of the MFM software, it may be able to make use of the proprietary fax machine features. For closed user group operations, this can add up to significant performance advantages. For example, the Okidata DOC*IT 4000 PS can send PostScript between like machines, and the Panafax UF-766 supports encryption. If you already have an established network of fax machines, such as for relay operation, then that may be another reason for considering an MFM over separate peripherals.

Need For a Fax Machine. Nearly all offices seem to need a "real" fax machine. It's hard to get away from the fact that everyone seems to have one regardless of whether they have fax modems, scanners, and printers. Somehow, there always seems to be the need to transmit paper documents and print out some, if not all, documents received by fax. If you consider that you will likely pay at least $400 for an office-quality fax, then the cost of an MFM seems even less. If you have to get a separate fax anyhow, why not get one that can do multiple jobs?

Multifunction Fax Software

All multifunction faxes require software that runs on the computer in order to take advantage of their multifunction features. It is this special MFM software that determines the visible personality and features of these machines. As with the MFM themselves, there are many software packages, and each has unique benefits suitable to different needs.

There are four packages that seem to represent diverse approaches toward using MFM. SciFAX from Cognition Sciences is a DOS-based program aimed at the more knowledgeable power fax user and desktop publishing and graphics computer user who is oriented toward maximum application of sophisticated MFM features and the production of high-quality fax images. L.A. Business Systems CAN-FAX is a DOS-based that is also available for use with Windows and will work on many local area networks. MacFacsimile for the Macintosh and UltraFAXit for Windows are from Vivitek, Inc. Both products run native in their respective environments.

SciFAX from Cognition Sciences, Inc. This package is oriented toward the user who is looking for maximum MFM performance and flexibility in use with DTP tools. SciFAX is built around the concept that the fax machine is an independent and intelligent unit connected to the computer and capable of operating on its own, even when the computer is turned off. The product aims to bring all the powerful features of a sophisticated MFM like the Canon L775 under computer control.

For the desktop publisher, SciFAX conveniently integrates with Ultrascript PC, GoScript, and Freedom of the Press PostScript to fax conversion utilities by providing transparent access to these products when using SciFAX to print, fax, or convert between file formats. The program masks the interaction with the conversion utility, so that when you tell SciFAX to send and print a PostScript document, it calls the appropriate conversion utility, creates a fax document on disk in the correct format, and uploads it to the fax for printing or faxing.

SciFAX handles compatibility with other programs through its ability to convert files in its native MMR-based fax format (.FAX) to and from files in the widely used PCX format. The conversion utility, Scifaxer, that comes with the package, is among the fastest conversion and compression utilities around, taking only two to three seconds per typical page on a 486/33 class machine. It and CAN-FAX are the only two packages that use MMR file compression which results in file sizes that are typically 30 percent smaller and take 30 percent less time to move between the MFM and the computer. Multipage documents are handled using a file naming convention that permits sequential processing. Sequential files have eight-character names, with the last three characters forming a numerical series starting with 001. For heavy DTP users or anyone planning to store or archive large numbers of documents, MMR compression translates directly into a 30 percent cost savings on storage.

For the high-power fax machine user. SciFAX provides a special form of relief. High-powered fax machines like the Canon L77x can be extremely complicated to use for tasks other than simple sending and receiving. Although a trained sales force makes it look easy, the truth is that using the advanced features of fax machines is difficult. Tasks like programming broadcast lists or storing documents in fax memory for later transmission must be done using a bizarre sequence of codes tapped out on a dizzying array of cryptically labeled buttons. To make matters worse, mistakes in button-punching usually are punished with a loud and insulting melt-down alarm that makes sure everyone knows "you screwed up!"

SciFAX translates the dizzy array of fax buttons into a rational, menu-driven computer interface. It does the dirty work, maintains broadcast lists, and controls scanning, printing, memory use, and the display of fax status information. The user can do everything at the computer that the trained fax salesman can do with the dizzy array of buttons.

For example, the Canon L77x machines have 99 electronic file folders, or EFFs, which allow storage of up to ninety-nine separate documents in fax memory. SciFAX lets the user control the contents of each EFF independently. As an added convenience and aid to remembering the contents of each numbered EFF, *SciFAX* maintains a list of these file folders along with brief one-line descriptions entered by the user. EFFs can be loaded from the computer, the fax (by receiving) or from the scanner, and can be sent from an EFF to printer or fax again using *SciFAX*. A *Canon L7xx* fax with the minimum configuration of 768k memory would have a limit of approximately fifteen typical pages using MMR compression. The amount of fax memory can be raised to a total of 3.75Mb which would store about seventy-five such typical pages.

In contrast with the other MFM software packages, SciFAX does not try to mask the function of the program or fax. Instead, it embraces this information and makes it more useful. It is conceivable that for some fax users, SciFAX would add sufficient value to justify the addition of a very-low cost computer to a high-power fax like the Canon L775 simply to serve as an interface device.

For the sophisticated user, *SciFAX* is an education. The documentation is the most thorough of any of the multifunction fax software programs, and it gives lots of examples and detailed technical explanations of what is going on and why. The creator of the product and author of the manual, Richard Larratt, has an encyclopedic knowledge of fax technology, and in conversation is more than happy to wax didactic. The manual is definitely in character. At just over two-hundred pages, the *SciFAX* manual is alone arguably worth the price of purchasing *Sci-FAX*.

CAN-FAX from L.A. Business Systems. CAN-FAX is more of a general-purpose program oriented toward business use. It does not allow access to EFFs or interface to PostScript conversion utilities like SciFAX does, but it has other handy features. Here is a summary of what CAN-FAX offers:

- Forms Overlay, for creating forms, letters and other documents that are repeatedly used but require variable data like a recipient name or other text.

- Mail-merge-style broadcasting from a word processing package that includes electronic generation of letterheads and signatures on outgoing faxes.

- A cut and Paste feature, for manipulating fax documents as bitmaps.

- Limited OCR for turning scanned documents or received faxes into text files. This includes the ability to work with third-party OCR software.

- Network Sending, which turns a PC into a network fax server that will handle fax transmission for any other computer on the network. Transmission is done by placing a file in a directory on the network. The NET-SEND program periodically checks this directory and sends any files it finds. Logging and control are done through separate ASCII command files or special commands embedded in the fax documents themselves.

- Automatic Routing of received faxes based on the use of a special cover page.

- PCL4 file compatibility in portrait mode. Allows direct printing of documents intended for Hewlett-Packard Laser Jet Series II printers. Will work with downloadable fonts including TrueType and Bitstream. This is in effect a PCL4 emulator driver.

- DXF and DC2 output from scanned documents. These file formats provide compatibility with a range of computer aided design (CAD) programs.

- PostScript file printing and faxing using GoScript as the conversion utility.

- Windows operation using a set of utilities that allow CAN-FAX to operate as a background task or as a Windows printer.

CAN-FAX shows its business orientation in how it masks some of the details of the fax machine from the user. For example, it handles printing of large documents that exceed the memory of the fax by automatically breaking the document into sections and sequentially uploading and printing the sections. It handles multipage scanning by automatically creating sequential file names for each page of a scanned document downloaded from the fax to the computer. This masking of details can be a drawback when faxing documents that exceed the size of the fax memory. Such documents would be broken up into multiple transmissions, which could be confusing to the recipient.

For those who use a Windows environment, the Windows Runner feature is strongly recommended because it lets you run CAN-FAX in the background without paying a major performance penalty. Running CAN-FAX in the background really takes advantage of the purpose for background processing. CAN-FAX can perform file conversions, uploads, and downloads from the fax machine, printing and faxing while you perform other work.

Note, that although it is possible, it is not desirable to attempt to run CAN-FAX in a DOS window as a background task. Operation in this manner results in very sluggish performance on the computer and a high possibility of errors during downloads. This is true even when using a relatively fast computer. Among the unpleasant side effects of doing this are excessive typing delays, jerky cursor response to the mouse, and missed or delayed mouse clicks. L.A. Business Systems recommends either using the Windows Runner for CAN-FAX or running CAN-FAX exclusively in a full-screen window.

The network sending feature of CAN-FAX would be most appropriate to a small to mid-size office that already has a Canon fax and computers on a network, or a larger office that has a $29,900 Canon FAX-L6500 or a

$17,995 FAX-L4600. With CAN-FAX, all network users could easily send documents by fax using the Canon machine. When the network sending feature is installed, users send faxes by placing documents or command files in specific network directories. CAN-FAX looks in these directories periodically for files to send out and processes any jobs it finds.

UltraFAXit from Vivitek, Inc. This package is among the simplest and easiest to use of the MFM software. It runs under Windows and behaves very much like the other print utility programs mentioned earlier in this chapter. In fact, UltraFAXit is a derivative of FaxIt, a program that is now called FaxWorks. Both have a copyright from Alien Computing and they have very similar interfaces and operation. If you install FaxWorks and then install UltraFAXit, the two programs get confused with each other sometimes resulting in strange behavior.

UltraFAXit provides printing, scanning, and faxing, but the interface completely masks the functions of the fax machine. There is no hint that EFFs exist, that the fax has memory, or that it is even a triple-function machine. Printing and faxing are differentiated only by a single setup toggle button that directs the output to printing or faxing. Scanning is simply a single item on one of the pull-down menus.

Faxing and printing with UltraFAXit are done through the standard Windows print procedure. To print or fax a document you select UltraFAXit on Canon and indicate what pages you want printed. If the fax/print toggle is set to fax, you will see a dialog box that asks you to enter the name, fax number and optional cover page information of the recipient. Documents that have been transmitted or printed are listed in the transmit log, where they can be viewed, rescheduled for printing or faxing, or deleted. The transmit log also contains status information.

Documents received by the fax machine are automatically routed to the computer, and appear as a listing in the receive log. Double-clicking on any of these entries gives the user options to view, delete, or check status. The viewer utility is Windows based, but does not have any controls for zooming or fitting the image into a frame. FaxWorks is considerably more flexible in displaying faxes and UltraFAXit may eventually inherit these abilities. Documents are scanned in using a pull down menu option. They are treated just like received faxes. They are loaded onto the computer and appear in the receive log.

Export of documents takes place through the viewer. The viewer uses a Save As menu option, and will save the received fax in PCX, DCX, TIF (uncompressed), TIF (with Group 3 compression), and Canon fax formats. Only the DCX format allows you to save a multipage document.

Sending and scanning functions allow the user to control resolution and time of transmission. There is a simple phone book for storing frequently used phone numbers, but no import or export method. The cover page is clean, attractive, and business-like, but it cannot be customized. All operation takes place in the background and is completely transparent except for a slight performance reduction in the computer. A fast machine should be able to handle this without any difficulty. The documentation is simple, straightforward, and concise. UltraFAXit is the only one of the three packages that comes with the cable required for connecting the computer to the MFM.

This is a simple program that quickly and with minimum time investment turns a triple-threat MFM into a high-utility computer peripheral. The degree to which the interface masks the details of fax operation seems to mitigate the delays and other side effects of the relatively slow serial link between the MFM and computer. The tradeoff is that there is limited access to the advanced features of the fax, like direct control over the EFF memory. For the workstation user who is looking for simple, immediate utility and is not concerned with using the more advanced fax machine features of MFMs, UltraFAXit is the easiest solution.

MacFacsimile from Vivitek, Inc. This product is different from UltraFAXit in that it is somewhat less transparent in its operation but provides more direct control over the fax machine and its features. The most noticeable difference between MacFacsimile and UltraFAXit is the explicit control it provides over the EFFs. A user at the computer can directly upload or download a document into any one of the ninety-nine EFFs or control printing, scanning and faxing using EFFs. Two very useful features are Show Fax Files and Clear Fax Files. Show Fax Files allows searching for EFFs that are not empty and Clear fax files is a fast way to clear one or more EFFs automatically. Unlike SciFAX, MacFacsimile does not keep any description of these storage locations. It requires the user to remember the contents of the numbered EFFs.

MacFacsimile installs as a print resource in the System Folder. It allows printing, faxing, and storing of documents as fax format files through the Print dialog box. Its operation is entirely consistent with the usual Macintosh print procedures. It will handle multipage documents, but the manual warns not to exceed the limit of the fax machine memory. When the application itself is running, it provides additional facilities that allow exporting of stored fax documents to the fax memory and explicit control over printing and faxing from memory. Printing of text files is made faster by exporting text instead of a fax image and using the built-in processing capabilities of the fax machine.

Scanning is directly controlled by the computer. Documents are first scanned into an EFF from which they can be printed, faxed, or imported into the computer. Import formats include PICT, TIFF, MacPaint, and Store formats. The Store format is used only for storing fax documents. Documents stored in this format cannot be viewed or edited. All scanning options for resolution, halftones, and lightness are controlled from the computer.

Fax reception is directed by a toggle either to the fax memory or to the fax printer. For a document to eventually make its way from fax reception to computer, it is first received into memory. When receipt has been detected, the operator can then import the document into the computer. Detection involves looking at the display on the fax for the message "Received into memory," and then using the Show Fax Files command to determine the EFF containing the received fax. The fax is then imported the same as it would be for scanning.

A simple phone book allows convenient storage of frequently dialed numbers. A special list processing feature allows the creation of text files containing long lists of fax numbers for use in broadcasting. A Print Receiver ID feature allows a string such as the name of the recipient to be printed at the top of each document. This can be quite useful in eliminating cover pages and solving the addressing problem, especially for large broadcasts. A simple viewing facility is provided for examining stored or imported documents. There are two zoom settings.

Regarding document preparation, the manual advises that any screen font can be used, but sending good-quality text requires either the installation of bitmap fonts at sizes three times larger than those actually used or the use of outline fonts like Adobe Type Manager or TrueType. MacFacsimile will also induce some stretch in documents in the interest of maximizing the quality of lettering when faxed. The result is that text will stretch out toward the right margin more than originally specified. A method for correcting this and producing documents with correct aspect ratios is given in the manual.

MacFacsimile will run under either System 6 or System 7. When available, it also makes use of Adobe Type Manager and TrueType fonts. It supports the Canon L770, 775, and 850 machines. A cable for connecting the fax to the Macintosh is provided.

Distortions in Digital Direct Fax Documents

Fax documents have built-in distortions that are by-products of the conversion or printing process. Most of these distortions are tolerable, and some of them are not even noticeable without close examination, but for situations in which dimensional accuracy is critical—such as producing artwork for advertisements or appearance-sensitive layouts for newsletters or product literature—it is useful to know what kinds of distortions happen and how to handle them.

Stretch

Stretch is a horizontal or vertical linear distortion. It evenly expands or shrinks an entire document. It usually happens both vertically and horizontally at the same time, but not in equal proportions. Since everything in the document is either reduced or enlarged in proportion, stretch is usually minor and thus not noticeable unless the absolute dimensions of a figure or the borders of a document are measured. For example, if you have a document with a frame around it that is intended for reproduction or insertion into an area of a specific size, a DDF of that document probably would not exactly match the required dimensions.

One situation where this was a problem was the creation of an advertisement for placement in a magazine. The artwork for the ad needed to fit exact dimensions, and because of time limits the ad was created using PageMaker and then transmitted to the publisher as a fine-mode digital fax using Winfax 3.0 and an Intel SatisFAXtion 200 modem. Once received, the fax was to have been reduced to fit the space for the ad. With the reduction, the effective resolution would be better than 300 dpi, masking the fact that the original had been a fax. When the art was proofed for content, it was also checked for size and was found to be incorrect. Checking the dimensions in the layout program against an actual fax of the document showed that there was a discrepancy between the dimensions as shown by the rulers on the DTP program and the actual result as printed at the fax.

Stretch seems to affect both fax print and conversion programs, but to varying degrees. Each program seems to have its own unique vertical and horizontal stretch. Table 5-2 gives the results of tests of several programs.

The test document was a 7-inch square drawn with a half-point rule using both Aldus PageMaker and Harvard Draw. There was no difference in the results generated by the two applications. All faxes were transmitted to a Panasonic PD-160E which printed the tests on thermal paper.

PROGRAM	SOURCE	OUTPUT DEVICE	RES.	VERT SIZE	HORIZ SIZE
Winfax 3.0	file/print	SatisFAXtion	fine/std	6.98	7.06
BitFax	file/print	SatisFAXtion	fine/std	7.12	6.86
Ultrascript	EPS file	Gammafax	fine/std	6.84	7.00
GoScript	EPS file	Gammafax	fine	6.96	7.04
UltraFAXit	file/print	Canon L770	fine	6.98	7.06

Table 5-2: Stretch Distortion From Fax Output Programs

This stretching or linear distortion may be created because these programs are only using an approximation for the actual resolution of fax. Fax is not *really* 203x98 (standard) or 203x196 (fine) resolution. Nor is it 100x200 or 200x200, as some programs will show. A look at the CCITT specification for fax reveals some interesting facts about how fax is defined.

Like any raster medium, fax is made up of pixels placed vertically and horizontally. While nearly every other raster image definition is made up of square pixels specified in integral quantities per unit of measure (for example, 300x300 dots per inch for laser printers), fax actually uses rectangular pixels specified in nonintegral horizontal and vertical resolutions, as shown in Figure 5-27. This converts into inches, as shown in Figure 5-28.

Vertical: 3.85 lines/mm (standard)
               7.7 lines/mm (fine resolution)
Horizontal: 1,728 dots on a 215mm line
               2,048 dots on a 255mm line
               2,432 dots on a 303mm line.

Figure 5-27: Fax Resolution as Specified by CCITT

Vertical: 97.79 dots/inch (standard) or
               195.58 dots/inch (fine resolution)
Horizontal: 204.145 dots per inch (using 1728 horizontal)
                   203.997 dots per inch (using 2048

horizontal)

                   203.871 dots per inch (using 2432

horizontal)

Figure 5-28: Fax Resolution Converted to Dots Per Inch

Note that the specification does not convert evenly into an integral number of dots per inch. Also note that the ratios of 1,728/215, 2,048/255, and 2,432/303 are not equal! As a result, depending on which conversion was chosen by the creator of the fax utility or conversion program, and on the degree of rounding used, each will give different results. Numbers in common use are 100x200, 200x200, 98x204, and 196x204.

Compensating For Stretch

In most applications, stretch should not be a problem, since most documents are not dimensionally critical. But when accurate dimensions are important, you will need to compensate for the effects of stretch.

Before correcting, it is necessary to determine the amount of stretch. A simple test to determine how much stretch or shrinkage you are getting is to draw a square box using a DTP program and print it to fax. Then, take a ruler and measure the actual dimensions. It is almost a sure bet that the scale will be off!

One can compensate for stretch by determining the exact proportions by which the horizontal and vertical dimensions are distorted and then scaling the entire document or selected objects within the document. For example, if you draw a 6 inch square and you get 5.875 inches on the horizontal side, you would compensate by drawing a figure that was 6 inches times the ratio of 6/5.875, or 1.0213. Thus you would draw a 6.128 inch line on the horizontal side.

Differences in Target Printers

Document layout can be affected by the differences among target printers. Specifically, what you print may not equal what you fax. Furthermore, what you fax with one program may not equal what you fax with another. For example, if you take a text-heavy document done in Microsoft Word or Aldus PageMaker and compose it for the Hewlett-Packard DeskJet 500 or the Windows PostScript Printer on FILE you will end up with a different result than if you had composed it for a fax print utility like Winfax 3.0 or FaxWorks. Differences will arise in horizontal and vertical scaling and most critically, in the spacing and quality of text.

Figures 5-29 and 5-30 illustrate this situation. The two figures were produced using the same document targeted for two different output devices.

Lorem ipsum dolor sit amet, consectetuer adipiscing elit, sed diam nonummy nibh euismod tincidunt ut laoreet dolore magna aliquam erat volutpat. Ut wisi enim ad minim veniam, quis nostrud exerci tation ullamcorper suscipit lobortis nisl ut aliquip ex ea commodo consequat. Duis autem vel eum iriure dolor in hendrerit in vulputate velit esse molestie consequat, vel illum dolore eu feugiat nulla facilisis at vero eros et accumsan et iusto odio dignissim qui blandit praesent luptatum zzril delenit augue duis dolore te feugait nulla facilisi.

Figure 5-29: Test Text Document Targeted to HP DeskJet 500

Lorem ipsum dolor sit amet, consectetuer adipiscing elit, sed diam nonummy nibh euismod tincidunt ut laoreet dolore magna aliquam erat volutpat. Ut wisi enim ad minim veniam, quis nostrud exerci tation ullamcorper suscipit lobortis nisl ut aliquip ex ea commodo consequat. Duis autem vel eum iriure dolor in hendrerit in vulputate velit esse molestie consequat, vel illum dolore eu feugiat nulla facilisis at vero eros et accumsan et iusto odio dignissim qui blandit praesent luptatum zzril delenit augue duis dolore te feugait nulla facilisi.

Figure 5-30: Test Text Document Targeted to Winfax 3.0

The vertical spacing (leading) is the same in both cases, but the set width of the lettering and possibly the track are noticeably different. With text flowed between fixed column guides, as would be the case with a newsletter, product description, or other more formally produced document, the difference is very noticeable. Other undesirable side effects stem from the same problem.

Page layout for long documents can be thrown off, causing page breaks to fall in undesirable places. When these differences appear, there are two choices. One is to redo the layout of the document to target it for the new printer or fax output device. This is acceptable when the document will not need to move around between output devices. But it is labor and time intensive, and can introduce new errors in finished copy. Usually, it is desirable for a single document to be able to print to a fax and to a printer without the layout changing. To do this, the best strategy is to try to use printers and fax utilities that are known to generate the same document The vertical spacing (leading) is the same in both cases, but the set width of the lettering and possibly the track are noticeably different. With text flowed between fixed column guides, as would be the case with a layout. For example, Ultrascript PC will generate the same spacing and results as the Hewlett-Packard DeskJet 500. A document composed and targeted for one should print with the same formatting and spacing as it would have on the other. Note, however, that the problem of stretch will still apply.

Aliasing

Aliasing is distortion due to differences in resolution between two raster image representations. It can happen when a raster image created for a 300 x 300 dpi laser printer is converted into a fine mode fax which is 203 x 196 dpi. Aliasing reduces the overall sharpness of details in fine text or drawings, and creates moire-like patterns in halftones. It can also cause problems with character uniformity, causing different instances of a character in a DDF document to look different.

Aliasing happens because when a conversion takes place between two raster image representations of different resolutions, there is no integral conversion between them. For instance, 300 x 300 does not convert evenly into 203 x 196. Therefore, when converted, the pixel boundaries of one image will fall inside the pixel boundaries of the other. This would not be true if the source image had horizontal and vertical resolutions that divided evenly into the target image resolutions. For example, 300 x 300 could convert with no aliasing into 600 x 600, because the conversion would simply make two pixels of the target image equal to every one pixel of the source image.

To get a better feel for this, consider the following exercise. You are given a source image of pixels as shown in Figure 5-31.

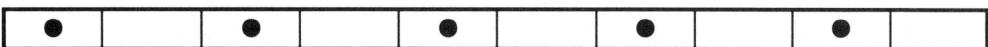

Figure 5-31: Aliasing Exercise Source Image

A black dot represents a black pixel, and a blank space represents a white pixel. The resolution is ten dots per unit. You are to convert it into the following two target images, shown in Figures 5-32 and 5-33, respectively.

Figure 5-32: Target Image Number 1

Figure 5-33: Target Image Number 2

Converting to Target Image 1 is easy. For each black pixel in the source image, you simply make the two directly corresponding pixels in the target black. The rest are left white. The result is shown in Figure 5-34.

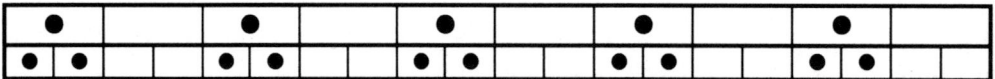

Figure 5-34: Result of Integral Conversion

There is no aliasing, because the borders of the source and target pixels correspond exactly—that is, they line up. But if we try this for the second target we get something like the image shown in Figure 5-35.

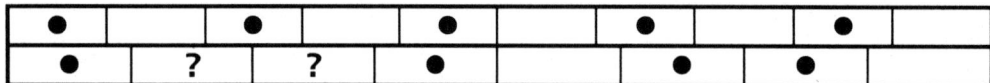

Figure 5-35: Result of Non-Integral Conversion

Look at the confusion that results. The new image hardly represents the old one, and in two cases we cannot tell whether to make a pixel black or white. Computers are no better at resolving these things than we are. When converting a 300 x 300 dpi image to a 203 x 196 dpi image, the ratio of conversion is approximately 3:2. This results in considerable aliasing. With large print or large featured graphics, this aliasing is much less noticeable. With fine-detail pictures or text, it can result in unacceptable degradation.

Figure 5-36 is an actual pixel-by-pixel representation of a 12 point lower-case letter "e" as rendered by Winfax 3.0 into a PCX file at 203 x 196. The matrix that forms the letter is quite coarse relative to its features. It is only 13 x 15 pixels. Converting to another resolution, such as 300 x 300, would be like converting this letter into a 19 x 22 matrix. If you tried to perform this conversion, you would see that on a line-by-line, pixel-by-pixel basis, you would encounter the same problem as in the above exercise.

Comparing Figure 5-36 to Figure 5-37 shows one result of converting the letter "e," and the effect of aliasing. The distortion is particularly noticeable in the lower half of the image. Note the apparent size shift is because the pixels in Figure 5-36 would be only two-thirds as large as the ones in Figure 5-37.

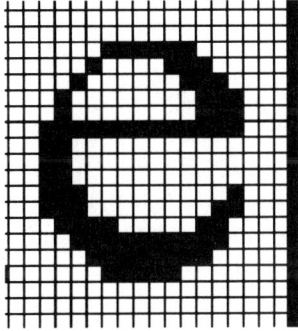

Figure 5-36: Actual DDF Fax Rendering of Letter "e" in 203 x 196 dpi

Figure 5-37: Pixel Map of Image from Figure 5-35
Converted to 300 x 300 dpi

Quantizing

Quantizing in DDF is similar to the quantizing that takes place in scanning that was described earlier in this chapter. Quantizing error in DDF affects only images that have features approaching the limits of fax resolution. It causes such features to either become obliterated entirely or to be fattened so that they come out larger. With DDF, the problems of skew-induced quantizing are eliminated, but the effect on small lettering and fine lines is similar. Using fine mode rather than standard mode resolution reduces the feature size at which quantizing will start to become noticeable, but only in the vertical direction.

Lettering. Quantizing usually causes letters to "thin out" and appear much "lighter" than they would if printed on a laser printer. Some fonts can wash out entirely and become gibberish. When designing documents it is best to try type samples in representative sizes by actually creating a DDF document with the sample type and transmitting it using the same processes anticipated for the final.

Fine Lines, Graphics, and Screens. The same applies to DDF here as for scanned input except that in the case of DDF the effects can actually be worse. An easy way to illustrate this is to create a document containing a perfectly horizontal hairline and then try to fax it in standard mode. It is highly unlikely that it will appear at all. When scanning, the skew error usually causes *something* to come out, even though the result would probably be quite jagged. Fine screens and graphics with fine detail will run into the same problem. A sufficiently fine screen may blank out entirely or be rendered as a pattern of dots with irregular placement.

CHAPTER 6

FAX PERFORMANCE

Image quality and transmission speed are the two key measures of facsimile performance. Many factors—how the image is created, the phone lines, the receiving equipment—can affect speed and quality. In this chapter we will examine these factors and discuss how you can control some of the variables.

Factors Affecting Transmission Time and Image Quality

Both sender and receiver have an influence over the transmission time of a fax. If you are doing fax broadcasting, then those factors that you as the sender can control are going to be more important. If, however, you are planning to call a fax-on-demand system regularly, then knowing what you can do to make fax transmit faster will be helpful, especially if you are using a one-call system and paying for the calls.

In most cases, image quality depends on the sender, since the way images are produced determines how they will appear to the receiver. The features present on the receiving machine, with the exception of ECM and special proprietary features like the *hyper smoothing* found in the Canon L775, should not affect the quality of the received image. In normal operating conditions and without any manufacturer-specific features in use, all faxes are capable of producing the same basic images. In general, fax images from the same sender printed by different machines should look the same. In fact, some inexpensive thermal

205

machines seem to produce sharper, clearer images than some very expensive plain paper machines.

The name of the fax performance game is how to down transmission time while improving image quality. For the most part, we are always trading off between these two. Increased resolution makes a better looking fax, but doubles transmission time. Use of halftone pictures can increase transmission times by more than a factor of ten compared to line drawings. The following sections discuss some of these tradeoffs.

Resolution

All Group 3–compatible fax machines and modems can transmit or receive in either fine or standard mode. Fine mode, at approximately 203x196 dots per inch, is twice the resolution of standard mode, which is approximately 203x98 dots per inch. The sender determines whether the transmission takes place in fine or standard mode before the transmission starts. A new super-fine resolution mode is available on some of the more expensive machines, but it is not in common use and is not supported by any of the present fax modems and software.

Fine mode makes documents look better, but takes twice as long to transmit. Obviously, if the image quality is acceptable, it is more economical to transmit everything in standard mode. Documents will transmit twice as fast and cost half as much. It is unnecessary to transmit ordinary typewritten documents in fine mode.

Users of fax machines should get into the habit of checking the resolution setting before transmitting. Many fax machines default to fine mode, and in most transmissions this is not necessary. Likewise, users of print utilities like Winfax or FaxWorks should know that fine is usually the default setting. If your fax use involves mainly sending typewritten or handwritten materials, checking and resetting the default resolution may be a quick way to immediately reduce your fax transmission cost by half without any noticeable difference in performance.

Resolution is determined differently for fax machines than for fax modems. With a fax machine, the operator determines the resolution of the transmission beforehand by pressing certain buttons on the machine. The machine then scans the document in either fine or standard mode during the transmission. The entire transmission takes place in that mode. Fax modems are different because they operate from stored images, and the resolution of a stored image, rather than the modem or software setting, determines the resolution at transmit time. For example, if you create a fine mode TIF file for transmission with a Gammafax CP fax modem, then it will always transmit in fine mode. If you took two TIF files, one standard and one fine, each would transmit at its own resolution.

DDF Documents. One of the main reasons to produce faxes digital direct is that DDF documents sent in standard mode can approach if not exceed the document quality of scanned faxes sent in fine mode. It is therefore possible to achieve both higher fax quality and faster transmission time using DDF documents. The following figures show this.

Figure 6-1 shows sample text as it would appear when scanned by a fax machine for transmission in standard mode. It is 8 point Arial normal. While it is readable if you know what to expect, enough details in the lettering are lost that one cannot be sure how the sixth word in the first line is spelled.

Figure 6-2 shows the same text as it would appear if scanned and transmitted in fine mode. It is quite clear and all the characters are properly formed, but now it would take twice as long to transmit.

To avoid the need to use fine mode to achieve high-quality, we can use DDF document production techniques as described in Chapter 5. Samples that illustrate the results are shown in Figure 6-3 and Figure 6-4. Figure 6-3 was

Lorem ipsum dolor sit amet, consectetuer adipiscing elit, sed diam nonummy nibh euismod tincidunt ut laoreet dolore magna aliquam erat volutpat. Ut wisi enim ad minim veniam, quis nostrud exerci tation ullamcorper suscipit lobortis nisl ut aliquip ex ea commodo consequat. Duis autem vel eum iriure dolor in hendrerit in vulputate velit esse molestie consequat, vel illum dolore eu feugiat nulla facilisis at vero eros et accumsan et iusto odio dignissim qui blandit praesent luptatum zzril delenit augue duis dolore te feugait nulla facilisi. Lorem ipsum dolor sit amet, consectetuer adipiscing elit, sed diam nonummy nibh euismod tincidunt ut laoreet dolore magna aliquam erat volutpat. Ut wisi enim ad minim veniam, quis nostrud exerci tation ullamcorper suscipit lobortis nisl ut aliquip ex ea commodo consequat. Duis autem vel eum iriure dolor in hendrerit in vulputate velit esse molestie consequat, vel illum dolore eu feugiat nulla facilisis at vero eros et accumsan et iusto odio dignissim qui blandit praesent luptatum zzril delenit augue duis dolore te feugait nulla facilisi. Nam liber tempor cum soluta nobis eleifend option congue nihil imperdiet doming id quod mazim placerat facer possim assum.

Figure 6-1: Sample Text, 8-Point, Scanned in Standard Mode

Lorem ipsum dolor sit amet, consectetuer adipiscing elit, sed diam nonummy nibh euismod tincidunt ut laoreet dolore magna aliquam erat volutpat. Ut wisi enim ad minim veniam, quis nostrud exerci tation ullamcorper suscipit lobortis nisl ut aliquip ex ea commodo consequat. Duis autem vel eum iriure dolor in hendrerit in vulputate velit esse molestie consequat, vel illum dolore eu feugiat nulla facilisis at vero eros et accumsan et iusto odio dignissim qui blandit praesent luptatum zzril delenit augue duis dolore te feugait nulla facilisi. Lorem ipsum dolor sit amet, consectetuer adipiscing elit, sed diam nonummy nibh euismod tincidunt ut laoreet dolore magna aliquam erat volutpat. Ut wisi enim ad minim veniam, quis nostrud exerci tation ullamcorper suscipit lobortis nisl ut aliquip ex ea commodo consequat. Duis autem vel eum iriure dolor in hendrerit in vulputate velit esse molestie consequat, vel illum dolore eu feugiat nulla facilisis at vero eros et accumsan et iusto odio dignissim qui blandit praesent luptatum zzril delenit augue duis dolore te feugait nulla facilisi. Nam liber tempor cum soluta nobis eleifend option congue nihil imperdiet doming id quod mazim placerat facer possim assum.

Figure 6-2: Sample Text, 8-Point, Scanned in Fine Mode

Lorem ipsum dolor sit amet, consectetuer adipiscing elit, sed diam nonummy nibh euismod tincidunt ut laoreet dolore magna aliquam erat volutpat. Ut wisi enim ad minim veniam, quis nostrud exerci tation ullamcorper suscipit lobortis nisl ut aliquip ex ea commodo consequat. Duis autem vel eum iriure dolor in hendrerit in vulputate velit esse molestie consequat, vel illum dolore eu feugiat nulla facilisis at vero eros et accumsan et iusto odio dignissim qui blandit praesent luptatum zzril delenit augue duis dolore te feugait nulla facilisi. Lorem ipsum dolor sit amet, consectetuer adipiscing elit, sed diam nonummy nibh euismod tincidunt ut laoreet dolore magna aliquam erat volutpat. Ut wisi enim ad minim veniam, quis nostrud exerci tation ullamcorper suscipit lobortis nisl ut aliquip ex ea commodo consequat. Duis autem vel eum iriure dolor in hendrerit in vulputate velit esse molestie consequat, vel illum dolore eu feugiat nulla facilisis at vero eros et accumsan et iusto odio dignissim qui blandit praesent luptatum zzril delenit augue duis dolore te feugait nulla facilisi. Nam liber tempor cum soluta nobis eleifend option congue nihil imperdiet doming id quod mazim placerat facer possim assum.

Figure 6-3: Sample Text in Standard Mode DDF Using Ultrascript PC

Lorem ipsum dolor sit amet, consectetuer adipiscing elit, sed diam nonummy nibh euismod tincidunt ut laoreet dolore magna aliquam erat volutpat. Ut wisi enim ad minim veniam, quis nostrud exerci tation ullamcorper suscipit lobortis nisl ut aliquip ex ea commodo consequat. Duis autem vel eum iriure dolor in hendrerit in vulputate velit esse molestie consequat, vel illum dolore eu feugiat nulla facilisis at vero eros et accumsan et iusto odio dignissim qui blandit praesent luptatum zzril delenit augue duis dolore te feugait nulla facilisi. Lorem ipsum dolor sit amet, consectetuer adipiscing elit, sed diam nonummy nibh euismod tincidunt ut laoreet dolore magna aliquam erat volutpat. Ut wisi enim ad minim veniam, quis nostrud exerci tation ullamcorper suscipit lobortis nisl ut aliquip ex ea commodo consequat. Duis autem vel eum iriure dolor in hendrerit in vulputate velit esse molestie consequat, vel illum dolore eu feugiat nulla facilisis at vero eros et accumsan et iusto odio dignissim qui blandit praesent luptatum zzril delenit augue duis dolore te feugait nulla facilisi. Nam liber tempor cum soluta nobis eleifend option congue nihil imperdiet doming id quod mazim placerat facer possim assum.

Figure 6-4: Sample Text in Standard Mode DDF Using FaxWorks

produced using the PostScript to fax conversion utility Ultrascript PC; Figure 6-4 was transmitted using FaxWorks Plus from SofNet. Both of these samples, while slightly different in their rendering of the individual characters, produce completely unambiguous copy. It is clear how each word is spelled.

Receiver Speed

Transmission speed is determined by both the sender and receiver. The two negotiate the speed at the start of a transmission and pick the slower of the two. Thus, if the sender is able to transmit at 9,600 bps and the receiver is only able to handle 4,800, the transmission takes place at 4,800. Most Group 3 transmissions take place at 9,600 bps. You should expect to encounter speeds other than 9,600 in fewer than 5 percent of all transmissions.

Compression

As previously mentioned, there are three kinds of compression techniques in widespread use. MH (modified huffman) is the most widely used but the lowest performance. MR (modified read) is an improvement over MH and is available on many fax machines. MMR (modified modified read) is the highest performance and will yield the lowest transmission times. It is typically only available on higher priced fax machines. For most typewritten or handwritten documents, MMR offers approximately a 30 percent faster transmission time when compared to MR. With the exception of the Gammafax CP, all fax boards seem to use only MH. The Gammafax CP is only able to receive, but not transmit, in MH, and it does not handle MMR at all. This seems to be an area where fax machine makers have stayed ahead of fax modem and modem software makers. Both sender and receiver must have the same capability for compression. If they differ, they will use the lowest common denominator.

Telephone Line Quality

Transmission speed is also affected by line conditions. Noise, inadequate bandwidth, and weak signal due to a poor connection will sometimes force a transmission to drop down ("fallback") to a lower speed. The Group 3 fallback speeds down from 9,600 bps are 7,200, 4,800, and 2,400 respectively. Poor line quality may also cause other transmission problems, such as scan line drop out. (See also the section on Error Correction Mode.)

Fallback is bad for two reasons. First, it slows down the overall transmission. If you are sending a long document, then you will pay a substantial penalty in telecommunications costs as the result of increased transmission time. Second, falling back takes time. The process by which the sender and receiver arrive at the conclusion to drop to a lower-than-maximum speed takes several seconds per try, and lower speeds are attempted sequentially in descending order. Thus, if a transmission starts out at 9,600 but has to drop down, it first goes to 7,200, then 4,800, and finally, on the third try, it would go to 2,400. Each try at "training" costs several seconds.

Fallback can happen in the middle of a transmission as well as in the beginning, although this is unlikely. If a transmission does fallback to a lower speed, it may not always be a good idea to continue at that speed, depending on the cause. If the cause is not related to the receiving fax (for example, it is a 4,800 bps machine) or is chronic, such as a connection that is always bad, then it is probably best to continue with the transmission. It would be unlikely that one could get a better connection and hence a faster speed. If, however, the normal condition is a 9,600 bps transmission and you are transmitting a long document and experience fallback, it may be wise to disconnect and retry the call. The cost of disconnecting and reconnecting for faster transmission is likely to be less than the cost of continuing the established call at a slower rate.

If you are doing a lot of transmissions—broadcasts, for instance—then you should be careful to monitor transmission speeds. What seems to operate well

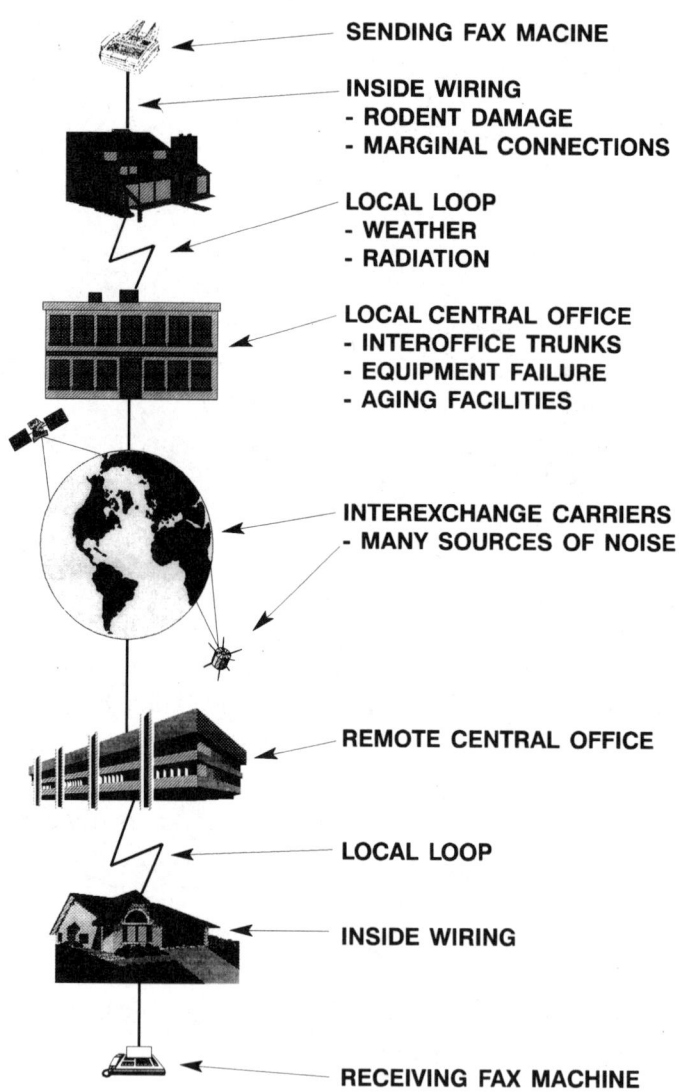

SENDING FAX MACINE

INSIDE WIRING
- RODENT DAMAGE
- MARGINAL CONNECTIONS

LOCAL LOOP
- WEATHER
- RADIATION

LOCAL CENTRAL OFFICE
- INTEROFFICE TRUNKS
- EQUIPMENT FAILURE
- AGING FACILITIES

INTEREXCHANGE CARRIERS
- MANY SOURCES OF NOISE

REMOTE CENTRAL OFFICE

LOCAL LOOP

INSIDE WIRING

RECEIVING FAX MACHINE

Figure 6-5: Potential Phone Network Problems

today can suddenly change. Reconfigurations in the telephone network have been known to cause unexpected changes in line quality which can have a severe negative impact on both transmission speed and received image quality. Some potential phone network problems are shown in Figure 6-5. In most fax broadcasting operations, transmission time is a major factor in profitability. Monitoring transmission results is therefore equivalent to minding the store.

Inadequate signal strength is another telephone line-related cause of transmission problems. It can come from several sources, but the most likely is poor performance in the local telephone line loop going between the subscriber premises and the telephone company central office. Copper lines have copper losses. If you have ever noticed while using the telephone that some connections seem much better than others, even within a local area, then you have likely experienced this variance in local loop quality. Some fax modems provide extensive monitoring and diagnostic information on transmission characteristics. Others provide features that allow compensation for poor lines. The Intel SatisFAXtion modems provide an adjustment for signal compensation that as it says in the manual, may need to be used when encountering poor line quality or inability to transmit. The Gammafax CP modems provide data on transmission speed, signal quality, line noise, and signal strength. This can be a real help in figuring out what is going wrong with large, routine fax broadcasts.

Another common set of problems with phone lines is caused by intermittent line noise, crosstalk, and signal degradation, particularly on long distance calls, though many sources can cause phone line problems. Here are just a few you may encounter.

Local Loop Problems. The lines that connect your office or home to the nearest central office of your local telephone company can have many problems. If you have overhead wires, squirrels can be particularly troublesome. They chew through cables and expose wiring to the elements. The symptom can be anything from low-level ticking noises to loud, periodic wailing or screeching. Local telephone company repair organizations are usually pretty good about responding to these types of problems.

Interoffice Problems. Central offices are interconnected with central office tie lines. These lines can sometimes go bad. The results can be insidious and the problems difficult to detect. For example, you can get seemingly perfect voice connections, but when fax or data calls are attempted, the results are poor transmission quality. They typically affect traffic from only one area, since calls must come over the specific interoffice line to have trouble. Telephone companies are very reluctant to even admit this kind of trouble is a possibility. They will claim the problem is yours and that they do not have any responsibility. Often, they don't feel any pressure to cooperate since they know you do not have any alternate source for service. If you encounter persistent problems with calls to or from a specific area, you may need to retain a telecommunications specialist familiar in dealing with telephone companies to help you handle it.

In one case in California, telephone calls from a specific area were constantly of poor quality. Even though the problem was easily reproduced, the local telephone company, Pacific Bell, consistently denied the existence of any problem. It was not until both the customers on both ends had spent months compiling evidence and escalating the matter in management, that the local telephone company finally admitted the problem was caused by a malfunctioning interoffice trunk and fixed it. (It helped that one of the two customers at the time was a telecommunications expert!) If you are involved in such a situation, plan to do your homework, and expect it to take several months, and possibly legal action, to resolve.

Long Distance Carrier Network Problems. Calls between the local telephone companies are carried by long distance carriers. These carriers operate national and worldwide networks to carry this traffic. The networks are composed of transmission facilities that can be changed to carry traffic differently depending on a number of factors including load, time of day, and cost. If you are transmitting lots of faxes using a long distance carrier, you may run into carrier-induced transmission problems, particularly if you transmit during the late night hours. Long distance carriers will sometimes reroute calls in their network over alternate routes. Their reasons may include maintenance and cost savings.

In a fax broadcasting operation in California, problems with transmissions late at night suddenly surfaced and became routine. These problems only happened with late night transmissions, and could not be duplicated at other times. The long distance carrier, AT&T, was willing to track down the problem with these cross country calls to a specific area of the U.S. It turned out that concurrent with the emergence of the fax transmission problems, AT&T had started a practice of reconfiguring the part of its network that served the troubled area. After some discussions about alternate carriers, the problem was fixed.

If you suspect your long distance carrier is causing problems, you can test this by rerouting your transmissions temporarily by using an equal access prefix code for another carrier. For example, if you are using AT&T and want to temporarily reroute calls over MCI or US Sprint, you can use access codes 10222 or 10333 respectively, before dialing the phone number. Likewise, if you are using MCI or US Sprint, you could reroute your calls over AT&T, you would use access code 10288. (For example, 10222-1-212-987-6543 would work from San Francisco to New York.)

PBX Switches and Extensions

If you place your fax broadcasting or response equipment on lines that are behind a PBX, be aware that the transmission bandwidth of some PBX products, especially those that claim to be all digital, may impose bandwidth or signal strength restrictions that have a negative impact on transmission speed. The symptom of this is that a fax will fail to transmit at full speed when connected behind a PBX but will transmit at full speed when connected directly to a telephone company provided trunk line. Telephone company provided phone lines already restrict the bandwidth of passable signals to 300-3,000 hz, and some PBX makers have seen fit to further restrict that in an attempt to achieve economy. The 9,600 bps modems—fax, data or otherwise—depend on having nearly all of this available bandwidth, and consequently suffer when it is the slightest bit impaired. The 14.4k bps modems are even more sensitive to this. If you know your fax equipment is connected behind a PBX, then you should make several tests by transmitting to other machines both near and far to be sure you are getting maximum transmission speed. One quick way to monitor this is to examine the transmission logs of either your fax machine or modem during routine use.

Error Correction Mode

Error correction mode (ECM) can improve overall efficiency of a transmission by helping to avoid the need to retransmit entire pages that may have partially unreadable images due to missing or erroneous scan lines caused by line noise. With ECM active during a transmission, scan lines are checked for errors. If errors are found, they are corrected. Without ECM, scan lines containing errors are "dropped," resulting in a fax that looks like it has thin horizontal slices missing. Scan lines with errors are usually caused by intermittent line noise, crosstalk, or signal degradation due to poor telephone call quality. (See the section on phone line quality earlier in this chapter.) Figure 6-6 is a somewhat

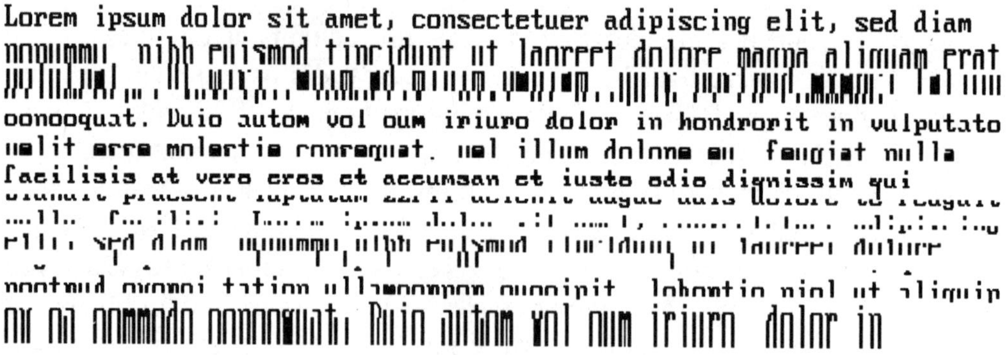

Figure 6-6: Scan Line Dropout

extreme example created for purposes of illustration. In most cases, scan line dropout would be much less severe.

While ECM provides the benefit of error free transmission, this comes at a price. Transmission times are lengthened depending on the amount of noise present on the line. The more the noise, the longer the transmission time. This is due to the retransmissions of groups of scan lines found to be in error by the receiver. Even under error free conditions, ECM will take slightly longer than standard transmission because of the overhead associated with the error correction protocol.

Image Content

The content of a fax has a great effect on how long it takes to transmit. Pictures, screens, and dense type will increase transmission time compared to line art, black or white spaces, and less dense type. The determining factor in transmission time with regard to image content is compression and how effective it is on a particular fax. The more black and white transitions, the less effective is compression. (See Chapter 3 for details on how fax works.) In some cases (such as pictures), compression actually works against fax, resulting in a net data *expansion*. Unfortunately, there is no provision in the Group 3 fax specification for turning off compression in such cases.

Pictures. Photographs or pictures like those in a newspaper or magazine article (screened photos) can dramatically increase transmission time. Any matter that has a continuous scale of lights and darks will result in slowing down a fax. If photographs must be incorporated into a document, try to minimize their size; the smaller the area of the picture, the less it will slow the transmission.

When possible, try to use line art instead of photos. Line art will appear sharp when sent digital direct and will dramatically reduce transmission time compared to a photo. For example, if you have product literature that shows a photograph of your product and you want to use that document in a fax

Figure 6-7: Line Art Picture—The Bearded Canadian

broadcast or place it in a two-call fax-on-demand system, have an artist convert the photo into an attractive line art drawing and use it in place of the picture.

A good example of line art is the cover page of the Intel FaxBACK in Chapter 2. It is contains an interesting, eye-catching graphic that is quite effective even in standard mode. Another example is shown in Figure 6-7; this line art drawing in fine mode that takes only thirty-one seconds to send.

Screens and Other Dense Patterns. Screens are also bad news for fax. A 50 percent screen is probably the worst thing you could do because it is the least compressible. To dramatize this, create a page of 50 percent screen and try faxing it using a fax machine. Watch as the paper goes through, and you will get a real feel for how screens can affect transmission time.

If you are designing new documents and know they will be faxed, avoid screens entirely if you can. If you must use them, use low-density screens. More white space, line art, and even completely black areas are better than even a 10 percent screen. Avoid any unnecessary lines, dots, and especially patterns. The less you put on the page, the faster it faxes.

If you are working with existing documents that were designed in the pre-fax era of business, try to get them redesigned if possible. Firms often use forms laden with screens in areas marked "For office use only." This is an inappropriate fax practice (and probably poor form design too), that needlessly increases transmission time. To highlight an area, use a bold border and a clear legend, or better yet, eliminate or minimize these areas in fax communications, and avoid transmitting anything that is not necessary. Another major source of transmission time waste are cover pages. It is amazing how many cover pages contain dense screens and other extraneous graphic detritus.

The Borland Tech Fax shown in Chapter 2 is a good example of where a screen should be eliminated. The Tech Fax logo on the cover page is filled with a screen. This screen probably adds an additional fifteen seconds to every fax response transmission, and communicates no useful information. The logo would look just as nice if it were strictly an outline; or it could be made more dramatic by use of all-black shadowing to give a relief effect while avoiding a screen. Tech Fax could also cut transmission costs in half by moving to standard mode for the entire transmission, instead of using fine mode. Both these measures taken together would more than cut in half cost of transmission.

Type. Large quantities of type like newsprint will take longer to transmit than smaller quantities of type of the sort used in a business letter. A good rule of thumb is that transmission time is directly proportional to the number of words on a page, and is somewhat independent of their size. If you have a document set in 10 or 12 point type, the difference in transmit speed will be marginal, or nonexistent; but if you add 10 percent more wording you can count on adding that much more transmit time. One tradeoff is type size versus resolution. Type size should be kept large enough that it faxes well using standard resolution and avoids the need to use fine resolution. Using DDF techniques, it is possible to use slightly smaller (and higher density) type sizes than with scanned documents.

White and Black Space. White space faxes faster than black space. Again, this is due to the way fax compresses images. The compression codes for all-black are longer than for all-white. As a conceptual simplification, transmission time can be said to be proportional to the amount of ink on the page. Since white is faster than black, you should avoid the use of black patches and reversed lettering, unless you're willing to pay the price for the appearance. Transmitting black takes less time than transmitting a screen, so if you need emphasis in a document, using black with reversed lettering is far less expensive than using a screen of the same area. Horizontal bands of white space fax fastest of all. If you have a choice in design to split a document, do so using horizontal white bands.

Colored Paper and Colored Areas. These can be as bad as screens or pictures, depending on the color and how dark it is. Very dark areas fax as straight black, which is not that bad. Certain colors, like a very light blue or faded green, fax as white and are effectively invisible. But other colors will create screen patterns as fax machines attempt to approximate their shading while transmitting them. Fax machines and scanners are very sensitive to papers that are reddish, pink, gray or brownish in color, but less sensitive to greens and blues. If you must use a colored paper or area in a document, again, try to minimize its size. When in doubt, check to see how it faxes first by actually transmitting a test.

Thin or Nonreflective Papers. Sometimes material like newsprint or documents printed on dark stock will fax very poorly. The result will appear to be letters on a gray or blotched background, and such an image will greatly increase transmission time while substantially decreasing readability. Thin paper can sometimes be pasted or taped to thicker paper to improve it for faxing. Nonreflective paper can sometimes be lightened using a copier before faxing. Be sure any result from a copier does not generate large gray regions, or those will fax as blotched screens and look terrible.

Dirt, Streaks, and Other Noise. Faxing dirty copy with streaks, dirt, bleed through due to thin paper, or extraneous images that arise from photocopying will slow down transmission. Some junk, like the notorious copier drive belt that gets included with a copy when the original is undersized, will produce a pattern very similar to a 50 percent screen. The result is that what should have been a one minute transmission is stretched into a three or four minute transmission. This is more common than you might think. Use clean copy when faxing.

Horizontal and Vertical Lines. A page of horizontal lines will fax faster than a page of vertical lines. Fax scans horizontally, so if a scan line is all black or all white it will transmit faster than if it is part black and part white. Obviously, a line of text has a lot of transitions due to the edges of each letter on the line, but the space between lines has none because it is all white. Many logos and other often-transmitted matter is composed of patterns of lines that run at some angle on the page. Horizontal lines, especially when done digital direct, are less costly in time than others.

Pictures and Grayscale Images

Even though they can be costly in terms of transmission time, sometimes it is desirable to fax a picture. There are several ways to do this, each of which gives a different quality result and has a different impact on transmission time. High-quality picture transmission is possible, and if it is necessary, the goal should be quality.

There are several ways to get pictures into faxes, including using fax machines, scanners, and MFM. One can even convert pictures from other digital formats. But in each case, picture quality is determined by how well a grayscale image, which has continuous varying shades of light and dark, is converted into a binary image, which is composed only of spots that are either black or white. There are several methods employed by fax machines and image processing software to convert grayscale images into binary halftones. They all try to represent the various shades of gray by varying the different patterns of black and white dots in the halftone. For example, a lighter shade of gray might be represented as a sparse pattern of black dots, and a darker shade of gray as a more dense pattern of black dots. The human eye takes these patterns and integrates them, allowing us to perceive levels of gray in what is really a just black and white picture.

Better quality fax machines and scanner software (for computers) will give better results because they will use more sophisticated techniques to represent grayscale. They will exploit the way the human eye perceives images in selecting the patterns of black and white to use to represent different kinds of gray. For fax documents, this means the quality of the resulting picture is determined by the sender. See Chapter 5 for a description of Canon UHQ and how this feature works on halftones.

Fax machines that have grayscale or halftone features can do an acceptable job of faxing pictures. Figure 6-8 shows what one of the more economical fax machines with halftone features can do. Compare this with a fax sent without using a halftone feature, as shown in Figure 6-9, and the difference is dramatic.

Figure 6-8: Halftone Fax on Panasonic PD-160E

Figure 6-9: Picture Sent Without Halftone Feature

Figure 6-10 shows a third example which was made using a Canon L770, and shows the kinds of halftone performance differences that exist among fax machines.

Most machines are set by default to not engage the halftone feature. This makes sense, since most fax transmissions consist of typed material and line art. It is not desirable to send text or line art using a halftone feature. When typed or other black and white material is sent using halftone, the lettering and other features that should be shown to be completely black end up grayed out, resulting in a less sharp, more diffused appearance, as well as in longer transmission time. (See Figure 6-11.)

The difficulty of producing a high-quality fax document that contains both sharp, full-black text and lines, as well as halftone pictures, is a good reason to use DTP tools and DDF techniques instead of scanning with a fax machine. With DTP tools, pictures that need to be placed in a document can be scanned separately using a computer-attached scanner like the Hewlett-Packard ScanJet. They are then converted to halftones and then included as binary raster images (bitmaps) in the document. When the document is finally turned into a fax, the text and line art are converted into black and white letters and lines, while the picture, which is already a scanned and converted binary image, is just "passed through" and onto the final fax. The difference is that the picture is scanned and converted to a halftone while the lettering and line art are not. Figure 6-12 is an example of such a document. The halftone imaging was done using the Canon L770.

Figure 6-10: Halftone Fax on Canon L770

Lorem ipsum dolor sit amet, consectetuer adipiscing elit, sed diam nonummy nibh euismod tincidunt ut laoreet dolore magna aliquam erat volutpat. Ut wisi enim ad minim veniam, quis nostrud exerci tation ullamcorper suscipit lobortis nisl ut aliquip ex ea commodo consequat.

Figure 6-11: Halftone Graying of Text

Beautiful Suburban Duplex Home

- **- Easy Access to Downtown**
- **- Plenty of Parking in Back**
- **- Can be Converted to Single Unit**
- **- Ideal for "in-law" Situation**

Offered Exclusively by
XYZZY Realty
415-555-1212
Please ask for:
Joe D. Realtor

Now you have a friend in the
Real Estate Business

Figure 6-12: DDF Document With Text and Halftone

Page Gaps

More pages cost more time, because each time a page is sent the sender and receiver spend time signaling the change of pages. If you have lots of information to fax, it is better to cram that information onto fewer pages with higher density than to try and send more pages with lower density. Fax in standard mode will render even 8 point type with adequate clarity for most purposes.

A natural idea would be to fax legal-size documents rather than letter-size documents. This would work and achieve the desired result of reducing the page count, but with one significant restriction. Many plain paper fax machines will try to print these legal-size pages on letter-size paper. The result will be a chopped up document at the receiving station.

Scan Line Delay

If you look through a fax transmission log you may notice that transmissions that take place at the same speed—say 9,600 bps—sometimes take significantly different times to transmit. One possible cause (other than compression, discussed previously), is scan line delay. This is a small delay imposed by the receiving machine on the sender. It is there to give the receiver time to advance the paper one scan line before receiving another scan line. Scan line delay can vary widely. The newest machines and most fax modems have zero scan line delay because they can buffer incoming transmissions and compensate for any mechanical movement that is necessary. Older or lower-cost machines do not do this, and, depending on their relative performance level, may impose delays between five and forty milliseconds per scan line. The most common for fax machines seems to be twenty milliseconds.

Document Design Tips

Here is a checklist of things to watch in fax document design and transmission performance.

- *Font size.* Use fonts of at least 10 points for sans-serif, and at least 12 points in size for serif.

- *Font technology.* When possible, use TrueType, Type 1 PostScript, or Adobe Type Manager.

- *Bitmap fonts.* When using bitmap fonts, install fonts that are at least three times larger than the largest font you intend to use.

- *Recommended fonts.* Fonts that are known to fax well include Lucida Sans, Stone Sans, Lucida Roman, and Arial, Helvetica and Gill Sans.

- *Font attributes.* If you want to select fonts, look for fonts that have unvarying and heavier lines, larger openings in letters, and relatively plain (sans serif) designs.

- *Proof with fax.* Don't rely on a printer to determine how a document will look. Proof your documents by actually faxing them using the same process you will for the final. Many fonts are "thinned out" when they are rendered for faxing versus printing. There can also be other distortions, as covered in Chapter 5.

- *Design for standard mode.* The temptation is to take the easy way out and design documents for fine mode since it is somewhat less demanding. If your documents are used occasionally, then this is probably a good tradeoff. If you are going to send a thousand copies of a document to a subscriber or customer list, designing for standard can give you quality with much better economy.

- *Design for letter size.* Most thermal paper machines can accept almost any size paper, but the increasingly popular plain paper machines mostly run on letter-size paper. Designing for anything larger than that will result in documents that are clipped-off by the receiver.

- *Watch margins.* Margins may not be consistent due to distortions. Watch them in the proof print.

- *Watch headers.* Some fax programs add headers to the top of each page. In some cases these can be suppressed; in others, they cannot. These headers can add anywhere from ¼-inch to ½-inch to the top of each page, increasing its length. The result is that you can end up with a document that is longer than your target size and ends up getting clipped-off at the receiver. The header can be very useful for routing. If you design this in, it can obviate the need for a cover sheet.

- *Use DDF.* Whenever possible, create documents DDF, as they will both look better and fax faster.

- *Monitor transmissions.* Watch transmission speeds by examining performance during transmission and looking at logs afterward. Fallbacks, disconnects, and other problems should be caught early.

- *Use your ear.* The human ear works very well as a phone line troubleshooting device. If you have trouble sending to a fax or receiving from one, call the receiving machine on a regular phone and listen for noise, distortion, echoes, or signs of other problems.

CHAPTER 7

FUTURE FAX TECHNOLOGY TRENDS

What in the world will happen to fax next?

In the mid-1980s fax seemed to go from obscurity to the center stage of business communications almost overnight. Quietly and without warning, fax technology snuck up on us. Fax had been around for a long time since its invention in the 1800s but it remained a virtually unknown and disregarded technology until the CCITT introduced the Group 3 fax recommendation. This seminal action seemed to magically elevate this former stepchild of electronic mail to a larger-than-life icon of popular culture. Suddenly, the media started carrying articles on fax and popular culture was buzzing with new catch phrases like "Can you fax it to me?" and advice on subjects such as "How to practice safe fax."

Now, less than a decade later and almost as unexpectedly, we have seen a sudden surge in computer-based fax technologies. Broadcasting, on-demand, and digital direct fax technologies are starting to hit the mainstream. Each day brings more new applications. And the agents of technological change seem to be working overtime.

Whereas in 1988 there was only one known way to convert desktop publishing documents to digital direct faxes, there are now dozens products that make this possible. In 1988 there were only a handful of fax modem manufacturers, but today the fax modem has become a standard feature of many packaged computers. The Fax Industry Guide published by the Institute for the Future lists over two hundred manufacturers of systems, boards, software, and services. Fax is today only just starting to take off.

In attempting to guess the future of fax technology, I have taken two views on what is most likely to happen. One view is of the near term, a timeframe of two to four years. The other view is of the longer term, perhaps ten years. The ordering of these two futures is more important than whether they actually happen in the estimated time frames. The time frames can vary but the longer term view, contrary to most forecasts, has been used to help predict the near term forecast, rather than the other way around. Let's take a look at the long term first.

Fax in the Digital Future

The dominant influence in the future of fax will not be from the world of computing or fax as we know it today, but from the land of telecommunications. ISDN, which stands for Integrated Systems Digital Networking, is rapidly becoming a reality. As this book is written, Bell Atlantic, one of the seven regional telephone operating companies, has just announced plans to begin the conversion of all telecommunications wiring in the state of New Jersey to fiber optic cable. This would includes homes as well as businesses.

This significance of this announcement is not that fiber optic cable is replacing copper wiring. It is that the conversion demonstrates the resolve of one of only seven companies that run the wiring of the U.S. to move quickly into the all-digital future of telecommunications, and make ISDN a reality.

ISDN translates into electronic, digital signaling, over phone lines rather than through analog signals. The word "ring" takes on a new meaning. Now, instead of a call just being indicated by a ring, it can be differentiated by what kind of ring it is. The new ring will tell us who is calling, what kind of call it is, whether a person should be alerted or a machine should handle it. In short, whatever is connected to the phone line will be able to decide how to handle the call before answering it. Thus, the problem of call differentiation—telling whether it's a fax, voice call, data call—will be solved, forming the switched network base for the building block of the telecommunications future.

That building block is a terminal device that combines all the functions we currently think of when we think of fax, phone, data terminal, and computer. In short, the computers of today will have become the generic communications and computing devices of tomorrow. What now is separate—phones, faxes, answering machines, computers, modems, terminals, switches and more—will be integrated into a single box.

The basis for this is already at hand. Such a product could exist now, given the current state of technology. There is plenty of evidence. PC-level cards exist that combine voice, fax, and data all in one unit and at incredibly low prices. The AT-bus computers have become generic engines used to drive everything from fax servers to PBX switches to voice mail systems, databases, and even entire networks. And interfacing to an all-digital medium would lower cost compared to the need to work with the analog signaling of today.

But the problems to be solved are not of hardware and software but of integration. One level of integration is between the terminal equipment (computers) and the public switched (phone) network. How do our computers differentiate between voice, data, and image traffic? ISDN is an enabling technology that will help solve this problem by providing a means of passing such identifying signals through the switched network. The terminal equipment and applications that drive it will need to do the rest. It may not be what the phone companies originally thought it should be, but in the end it will prove to be the technology that enables network-terminal integration.

Another level of integration is at the application level. What kind of applications could handle mixed modes of interaction? How will they work and what will they look like? We will not end up with countless ways of doing the same thing as we have in the past. Integrating functions requires dramatic simplification of how such functions are used, and therefore there will be standards for usage.

Open standards, de facto or de jure, will eventually emerge to make every aspect of such products—from the human interface to the wire-side hardware—compatible, interchangeable and integrated. We will therefore be able to deal seamlessly with "traffic" or "messages" that we previously subdivided into the categories of voice, fax, data, or other. Our universal terminal device will be able to handle them all with an interface that will be pushbutton-simple. And it will be intelligent enough to handle them automatically if that is what we want it to do. Translation: one machine to do fax broadcast, on-demand, voice messaging, data service, "multimedia," and perhaps even video. ISDN will enable the market and consumers of products and services will demand integration.

A critical part of ISDN is a public network technology known as Signaling System 7 or SS7 to those in the trade. This is the foundation that will enable ISDN to happen, and it is already widely deployed. It will allow the passing of information *about* a call to pass separately from the actual data or signal stream of the call itself, and is a key to passing this information to the terminal equipment (computers) that will need to differentiate calls in order to handle or route them. By the year 2000 it should be fully deployed in North America, and the digital, end-to-end, switched network call will be a reality. The digital circle will be unbroken.

Computing will follow these developments in telecommunications. This is not a technological phenomenon but rather a political one. Change in the world of computing is relatively free to happen without legislation or consent, because the level of investment required to introduce change is proportionately much lower than it is in the world of telecommunications. It's easier to decide to build an integrated voice, data, fax, or video add-on card; introduce a new operating system; or even invent a completely new computing architecture than it is to restructure the basis of the switched telephone network. The latter requires a lot of money, and, since it is regarded as a quasi-public issue, it also requires the will of the people.

So what will happen in computing? Follow in a frenzy it will. We will see smaller, less expensive, more powerful machines than we can even imagine today. (Of course. Could you imagine slower, larger, more expensive?) But the key difference will be integration. Applications, operating systems, and machines will have to meld together into a unified, consistent entity in order to make best use of what the new technological environment has to offer. The new machines will make today's operating systems and application programs look like erector sets built by Rube Goldberg himself. The new machines will more resemble the automobile than the computer of today. The prime fantasy of the computer counterculture—the making the computer into a wet-dream sex-charged super-car—will finally be in hand.

Computers as used by end users (versus as embedded circuits or subsystems) will have become the universal instrument (UI), able not only to arrange bits on a page or shovel them out a phone line, but also to distinguish communications traffic by type, process or handle requests autonomously, and yes, still interface with human eyes, ears, and mouth. They will replace the handset and keypad, the separate phone on your desk. When you send or receive a voice call or message, fax, file, or whatever traffic, your UI will handle it for you.

Suppose you want to call someone. You tell your UI to make a call. If that person is near a UI or has one in his or her pocket, you get to chat. During such a chat you decide you want to compare notes or exchange information. No problem. No need to hang up or even have a separate line. Just tell your UI you want the other person to see what you are seeing or have access to some information you have, which could be anything from a fax, to a data file, to a voice message. The boundaries between various kinds of information will blur.

But suppose you call and no one answers. Perhaps you get a recording. Maybe you receive a fax or file. Maybe you leave a voice message. Or a fax. Or just press a button that says "call me back" and automatically records your callback information. (Will this finally be the death of the pink message slip?) Maybe you will rummage through the information in your friend's UI because it recognizes you and you can find what you need without having to leave a message. Better yet, you didn't have to ask to chat, since you told your UI in the first place what you wanted and it just went out across the network to the other person's UI and got it. Machines will do the grunt work and conversation will end up reserved for interaction between people.

Fax as we know it today will cease to exist. Instead of being a separate technology that mainly involves the scanning and printing of paper documents, fax will have melded or expanded into the notion of imaging and image management. And images will be just one part of documents, which by then will have become multimedia-like, incorporating any form of communication, human- or machine-readable. Those who insist on connecting "antique" fax equipment to the network will be able to do so through small, intelligent interface boxes that act as translators between the all-digital present and the remote analog past.

The biggest change of all will be the way people conduct business with these new machines and the new network. Business will be increasingly transaction-oriented, and computer-based servers will be able to perform much more of the workload that is now associated with information handling. Record-keeping and retrieval, information exchange via forms, routine requests for information, authorizations, authentications, filings, invoices, payments, and all the rest of the input/output and bookkeeping work of society will happen between machines, thanks to an important enabling factor: authentication technology and policy that will signal that our society is finally ready to accept digital documents as legal.

The Near Term

OK, so much for the day *after* tomorrow. Now what about tomorrow? The near term trends in fax technology will be shaped by the current forces in the market, which we can see clearly from where we are today, as well as on the longer term future, which we just examined. The near term fax future is a stepping stone or transition period between what we have today and what is expected the day after tomorrow. Let's look at some specifics.

Transmission speeds of 14.4k bps will be the next standard. Some believe speeds using a yet-to-be-determined specification temporarily named V.Fast may leapfrog 14.4. My guess is that both will be offered in the same products, as the chip industry tries to standardize circuit specifications. Many products are already moving toward, or have, 14.4. Business users will be the first to move to 14.4k bps, resulting in a dumping of 9,600 bps machines on the home and home office market. Look for deep discounts in the coming years.

These deep discounts will drive lots of people to get fax who never before might have been considered consumers for such products. One notable new fax consumer: school age children under the legal driving age. What is the application? Try homework. Guess what junior wants for a birthday present. A fax, with its own phone line or a line sharing device. The North American Numbering Plan will become increasingly populated. Look for more area codes coming to a neighborhood near you.

Along with these higher transmission speeds will come much wider use of MMR encoding and an increase of products that will be upgradeable to Group 4 compatibility. The cost of building in circuits that can handle switched 56k bps service (the communications service type for Group 4 machines) is lower than that of building modems, given equal volume of manufacture. The standard modem in two years will have at least 14.4k bps speed and MMR encoding. Optional equipment will be a 56k bps add-on interface for Group 4, though Group 4 will not become widespread until ISDN is widely deployed to the desktop.

The vexing problems of how to handle inbound routing of faxes will continue to be a problem and will become increasingly annoying over the next two years as fax servers for LANs become very popular. People will love LAN fax servers for the convenience they provide in transmitting faxes from computers. Many solutions will be tried, all with limited success. Look for DID to become the routing solution for most large corporations. The fundamental problem will be in working with the "great unwashed masses" who will still write out cover pages by hand for routing information. They will simply remain unchangeable.

Unfortunately, we will not yet see full-page fax displays, and most people will still want their faxes printed. Even when full-page fax displays do become a reality, paper will remain very popular. Blocking on inbound lines will not be tolerated as well as queuing on outbound lines. Most inbound faxes will still end up printed. Separate lines for inbound and outbound traffic with specialized equipment will likely be the dominant strategy for fax transmission and reception on LANs.

OCR will continue to generate interest, but will not produce acceptable results for converting ordinary faxes into computer documents without human editing. The general purpose OCR problem with fax will not yield results anywhere near the level necessary to be considered reliable enough for communication. One notable exception: OCR will work for documents faxed digital direct. There, the relatively distortion free faxes will achieve 99 percent accuracy or better in conversion of DDF images to text or other document formats. This may be a boon to makers of OCR software, and permit digital direct fax to act as a least common denominator of machine- and human-readable document exchange. The implication is that fax may become a form of universal EDI (electronic data interchange). By being readable by both people and machines, DDF may be adopted as a way of linking otherwise incompatible manual and automated work processes, particularly between organizations. Documents will be exchanged where a sender faxes a DDF document and the receiver uses either a person or OCR technology to convert it into a computer document. When carefully managed, existing products are capable of 100 percent recognition accuracy given clean DDF documents. This may also become one way to handle routing on networks. Another possible boost for OCR: multifunction peripherals that include scanners. Local scanning dramatically improves OCR success rates.

Fax modem interfaces will not require standardization. CAS, Class 1 and 2, Faxbios, and T.611 will coexist, and vendors of products such as fax software will simply respond by expanding the number and type of products their products support. Most of this detail will be gradually hidden from the user.

Multiline operation will become increasingly commonplace. An awareness of this need has already started with LAN users, but will spread rapidly to non-LAN users in small offices and home offices. Fax software vendors in the near term will need to add multiline capability to their products. Many fax modems can already support multiline (per machine) operation, but most fax software does not.

We will see increasing acceptance and demand for a multifunction computer-fax-printer-scanner peripheral (MFM) as more of the computer world moves toward DTP and GUI (graphic user interface) environments and begins to accept the dominance of fax as the universal method for printed and transmitted communications. The multifunction peripheral will come from the same people who brought us the laser printer and the fax: large Japanese multinationals like Canon and one or two U.S. firms, such as computer peripheral giant Hewlett-Packard Company. Scanners will no longer be an afterthought for computers. Everyone will want one as users become accustomed to having them available as part of the office fax machine.

Multifunction software that handles voice, fax, data, and voice and fax response will begin to remove the boundaries between voice and fax equipment. These packages will handle both even better than do the systems of today. They will be supported by board-level products based on the PCMCIA (Personal Computer Memory Card International Association) standard. The popularity of using data for interaction will still be a distant third in comparison to voice and fax. Data will be used mainly by large firms can afford to support specialized applications and services. A major challenge will still be how to differentiate voice, fax, and data traffic on the same line. Some proprietary implementations for doing this may become de facto standards.

Base station computers with fax modems will be increasingly used to support fax-equipped laptops. Desktop workstations will evolve into base-station-and-laptop combinations, with fax-based modems as the connecting link. The model of interaction between the two will be similar to the current client-server concept. The base station will be the main point of communications and service for a roaming laptop. The laptop will just "check in" periodically to pick up voice, fax, and data messages.

Finally, look for an expansion of fax-based services—everything from the corner restaurant offering the menu of the day by fax-on-demand to large corporations turning to fax for customer service, and entrepreneurs offering specialized information services and highly targeted niche publications. As word of what can be done with the new computer-driven fax technologies spreads, everyone will start to get into the act. Your need will be someone's specialty. And there will likely be a fax product or service just for you.

Now Over to You

What are your opinions on the subject of fax technology? Do you know of upcoming technology that you believe will be interesting? Are you doing something with fax that you think would be important for others learn? Is there something you would like to see happen in the future of fax? I invite your discussion; fax me at 415-776-0615. In addition to your opinions, tell me something about yourself, what you do, and why you are interested in fax. I'll take the best comments I receive and try and quote or summarize them and make the results available by fax, of course. Be sure to include your return fax number.

Fax Resource Directory

This resource directory is organized into listings by category followed by a complete listing by name. Abbreviations are used for fax broadcasting (FB), fax on demand (FOD), desktop publishing (DTP), and direct dial (dd).

Sources of Information

These firms provide directories, multiclient studies, books, technical publications or other kinds of information.

American Facsimile Association
Artech House, Inc.
BIS Strategic Decisions
Buyer's Labs, Inc.
Fax Buyer's Guide
Fax on Demand Services Directory
Institute for the Future
Omnicom

Fax Broadcasting Products

Audiofax
Copia International

Dynacor Graphics
FaxBACK, Inc.
Ibex Technologies, Inc.
Information Transfer Systems
SpectraFax Corp.
VoiceLink, Inc.

Fax Broadcasting Service Bureaus

AT&T Enhanced Fax
Accelerated Voice
Brite Voice Systems
Cable & Wireless
Dynamic Fax
MCI
Swift Global Communications
Xpedite Systems, Inc.

Fax-On-Demand Products

ABS Systems, Inc.
Accelerated Voice
Audiofax
Brite Voice Systems
Brooktrout Technology
Copia International
DBC Associates
DemoSource
Document Retrievel Services
Dynamic Fax
FaxBACK, Inc.
Ibex Technologies, Inc.

Instant Information
Nuntius Corporation
Open+Voice Inc.
SpectraFax Corp.
Speech Soft
TRT (Telephone Response Technologies)
VoiceFAX Information Systems
VoiceLink, Inc.
Xerox Corporation
Xerox PaperWorks Support

Fax-On-Demand Service Bureaus

Accelerated Voice
Brite Voice Systems
Document Retrievel Services
Dynamic Fax
Instant Information

Desktop Publishing Products and Services

Canon USA, Inc.
Dynacor Graphics
Gammalink
Hewlett-Packard (HP)
Inset Systems, Inc.
LaserGo, Inc.
National Semiconductor
PM Ware

Fax Modems

Audiofax
Brooktrout Technology
Calculus
Everex Systems, Inc.
Gammalink
Global Village Communication, Inc.
Hayes Microcomputer Products, Inc.
Intel PCEO
National Semiconductor
Prometheus Products, Inc.
Rockwell International
SpectraFax Corp.
The Complete PC
Xerox Corporation
Zoltrix, Inc.
Zoom Telephonics, Inc.

Miscellaneous Products

Qualitas
Tall Tree Systems
VSI Telecom
WideCom Group

Complete Listing by Name

ABS Systems, Inc.
2500 Shames Drive, Westbury, NY 11590
Tel: 516-333-7900
Fax: 516-333-7953
Makes TAXX voice processing system with FOD feature. Does auto attendant, voice mail, IVR, and FOD.

AT&T Enhanced Fax
5000 Hadley Road, Room 1B13, South Plainfield, NJ 07080
Tel: 800-248-3329 x1000, 908-221-2000
Fax: 800-843-7467
Fax broadcasting service with nodes in Japan and U.K. Also has fax mail.

Accelerated Voice
25 Stillman Street, Suite 200, San Francisco, CA 94107
Tel: 800-799-0027, 415-543-7998
Fax: 415-543-6398
This serice bureau provides fax retrieval and broadcasting.

Alien Computing
38733 Ninth Street East, Unit R, Palmdale, CA 93550
Tel: 800-336-1166, 806-947-1310
Fax: 806-947-1387
FaxIt, Windows printer driver.

American Facsimile Association
PO Box 8748, Philadelphia, PA   19101
Tel: 216-963-9110
Fax: 216-963-9108
Publishes weekly Fax Focus newsletter on industry news and events.

Artech House, Inc.
685 Canton Street, Norwood, MA 02062
Tel: 800-226-9977 x4002; 617-769-9759
Fax: 617-769-6334
Publishes an excellent book for the engineer: *FAX: Digital Facsimile
Technology and Applications.* Highly technical and detailed.

Audiofax
2000 Powers Ferry Road, Suite 200, Marietta, GA 30067
Tel: 800-283-4632, 404-933-7600, 618-4241(dd)
Fax: 404-933-7606
Fax mail, fax on demand, guaranteed fax transmissions, and confidential fax
with receipt reports. Comarkets with Voice Pro, Telecorp, Baker
Audio/Telecom DialPro, and Dial Connections.

BIS Strategic Decisions
One Longwater Circle, Norwell, MA 02061
Tel: 617-982-9500
Fax: 617-878-6650
Industry analysis, multiclient studies, conferences on fax and other technologies.
Offices worldwide.

BIT Software, Inc.
830 Hillview Court, Suite 160, Milpitas, CA 95035
Tel: 510-490-2928
Fax: 510-490-9490
Makes Bitfax fax software for Windows and DOS environments. Has optional
OCR capability. Includes program for data modems/terminal emulation.

Brite Voice Systems

7309 East 21st Street North, Wichita, KS 67206-1083

Tel: 316-652-6500

Fax: 316-652-6800

Info on sports, Wall Street Journal articles, stock quotes, games, and other information update services.

Brooktrout Technology

144 Gould Street, Needham, MA 02192

Tel: 800-333-5274 demo/info, 617-449-4100

Fax: 617-449-9009

FlashFax multiline FOD system and TR-series fax modems.

Buyer's Labs Inc

20 Railroad Ave, Hackensack, NJ 07601

Tel: 201-488-0404

Fax: 201-488-1839

Extensive reports on fax products including machines. Use mainly by corporate buying staffs.

Cable & Wireless

1919 Gallows Road, Vienna, VA 22182

Tel: 800-486-8686, 703-734-7148

Fax: 800-486-7109

SureFax U.S. and Int'l fax broadcast service and fax mailbox service accepts input using DTMF/fax machine, PC, fax only and special 700 numbers.

Calculus

1761 West Hillsboro Blvd, Deerfield Beach, FL 33442-1530

Tel: 306-481-2334

Fax: 306-481-1866

LAN fax software and modems; EZ-Fax, EZ-FaxIt, EX-FaxLAN, EZ-FaxLAN ESS, EZ-FaxLAN ESL.

Calera Recognition Systems
475 Potrero, Sunnyvale, CA 94086
Tel: 408-720-8300
Fax: 408-720-1330
Develops OCR software to faxes into ASCII files.

Canon USA, Inc.
One Canon Plaza, Lake Success, NY 11042-1113
Tel: 516-488-6700
Makes high end multifunction fax machine equipment.

Cognition Sciences Corporation
RR#1, Demorestville, Ontario, Canada K0K1WO
Tel: 613-476-4003
Fax: 613-476-4406
Makes SciFAX software for multifunction fax machines.

Copia International
1964 Richton Drive, Wheaton, IL 60187
Tel: 708-682-8898
Fax: 708-666-9841
FaxFacts, FOD with 4 phone lines per machine. Demo line 708-924-7465 is
for Document Retrieval Services run by Digital Services Corp., a Copia user.

DBC Associates
1 Daniel Burnham Court, Suite 809, San Francisco, CA 94109
Tel: 800-329-4537 (sales), 415-421-6000 (int'l sales), 415-776-6227 (other)
Fax: 415-776-0615
$1,995 MessagePost FOD system. Ready to run, no programming needed.
Demo: 415-771-0870.

Delrina Technology, Inc.
895 Don Mills road, 500-2 Park Centre, Toronto, Ontario, Canada M3C 1W3
Tel: 416-441-3676 x352
Fax: 416-441-0333
Winfax fax software for Windows.

DemoSource
8345 Reseba Boulevard, Suite 202, Northridge, CA 91324
Tel: 800-283-4759
Fax: 818-718-9579
Big Mouth, a voice/fax system; IVR system is the software phonebook;
FaxTalker 9,600 bps PC-fax card and software.

Document Retrievel Services
644 West North Pratt Avenue, Schaumburg, IL 61093
Tel: 708-924-6464, 708-351-2266 (dd)
FOD service bureau; Startup cost $100, Monthly $19.95 + .25/page; 800,900
services; volume pricing available.

Dynacor Graphics
38 Galt Avenue, Toronto, Ontario, Canada M4M 2Z1
Tel: 416-461-4343
Fax: 416-461-8641
Make FAXMERGE, database-driven "mail merge" for fax distribution.
FAXPRINT, FAXOVERLAY, FAXFONT.

Dynamic Fax
5301 East 8th Street, Suite 305, Rockford, IL 61108
Tel: 816-398-9009
Provides 900 voice/fax, credit card capture, 900/800/local and fax broadcast
services.

Enhanced Systems and Consulting
5435 Executive Place, Suite A, Jackson, MS 39206
Tel: 601-982-2757
Fax: 601-932-2101
Software development house that tailors fax and email systems.

Everex Systems, Inc.
901 Page Avenue, Fremont, CA 94538
Tel: 800-821-0806; 415-498-1111; 800-458-3839
Fax: 510-683-2025 Mktg
Computers and peripherals, modems, graphic boards, network products,
EverFax 24/96 fax and data boardfor IBM PC/XT/AT.

Fax Buyer's Guide
Tel: 212-807-8220
Magazine reviewing fax machines and related products.

Fax on Demand Services Directory
Tel: 708-666-0722
Directory of FOD numbers and services offered for businesses and consumers.
For listings call: 708-924-3022.

FaxBACK, Inc.
15250 NW Greenbrier Parkway, Beaverton, OR 97006
Tel: 800-873-8753, 503-646-1114
Fax: 503-646-4999
FaxBACK FOD and broadcast equipment. FaxBACK 1.4, IVR available in
English, French, Spanish.

Gammalink
1314 Cheasapeake Terrace, Sunnyvale, CA 94089
Tel: 408-744-1430
Fax: 408-744-1549
Fax modems and associated software; Also EPS to fax conversion utility. FOD
info number: 408-734-9906.

Global Village Communication, Inc.
1204 O'Brien Drive, Menlo Park, CA 94025
Tel: 415-329-0700
Fax: 415-327-3808
Teleport, data/fax modem for the Mac.

Hayes Microcomputer Products, Inc.
705 Westech Drive, Norcross, GA 30092
Tel: 404-449-8791
JT fax board for PCs, PS/2; JT Fax 9600B for fax/data.

Hewlett Packard (HP)
700 71st Avenue, Greeley, CO 80634
Tel: 800-752-0900 x2211
Scanners, plain paper fax machines useful for DTP to fax applications.

Ibex Technologies, Inc.
847 Pacific Street, Suite C, Placerville, CA 95667
Tel: 916-621-4342
Fax: 916-621-2004
FactsLine IVR/FOD

Information Transfer Systems
8556 Katy Freeway, Suite 124, Houston, TX 77024
Tel: 713-466-3884
Fax: 713-466-3885
Make Subfile, BulkFax, LMFAXP, a program to increase speed of printing Gammafax fax files.

Inset Systems, Inc.
71 Commerce Drive, Brookfield, CT 06804
Tel: 203-740-2400; 800-374-6738
Fax: 203-776-5634
Makes Hijaak graphic conversion utility program which converts between almost all known graphic file formats including over a dozen fax formats.

Instant Information
66 Long Wharf, Boston, MA 02110
Tel: 800-666-9335, 617-523-7636
Fax: 617-723-6522
Large FOD service bureau providing end-to-end service. Over 5 years in business as of 1993; does work for *Consumer Reports, Inc. Magazine*, and *Business Wire*.

Institute for the Future
2740 Sand Hill Road, Menlo Park, CA 94025
Tel: 415-854-6322
Fax: 415-854-7850
Produces annual multiclient listing on fax industry products and services available.

Intel PCEO
5200 Northeast Elam Young Parkway, Hillsboro, OR 97124-6497
Tel: 800-538-3373, 503-629-7532
Fax: 800-458-6231 End-User Support
Connection CoProcessor, SatisFAXtion fax modems, related software. FaxBACK IVR system developed but marketed by FaxBACK, Inc.

L.A. Business Systems
497 Pleasant Ave, Highland Park, IL 60035
Tel: 708-433-6477
Fax: 708-433-7676
Makes software that drives multifunction fax machines from Canon. Has extensive network features and operation from Windows.

Laser Arts
89 Fair Street, Guilford, CT 06437
Tel: 203-458-9935
Fax: 203-458-2535
Professional quality fax image processing and typesetting.

LaserGo, Inc.
9369 Carroll Park Drive, Suite A, San Diego, CA 92121
Tel: 619-450-4600
Fax: 619-450-9334
Go Script software for PostScript to fax (and other graphical format) conversion.

MCI
113 19th Street, N.W., Washington, DC 20036
Tel: 202-872-1600
Dedicated fax network that offers broadcast, store & forward, mailbox, and text-to-fax conversion services.

National Semiconductor
2900 Semiconductor Drive, Santa Clara, CA 95050
Tel: 408-721-5000
Fax: 408-739-9803
QuadFax-96/QuadFax-96m board; QuadScript PostScript; QuadFax Programmers Tool Kit; Etherfax PC LAN system compatible with Novell Netware.

Nuntius Corporation
1904 Merrill Drive, St. Charles, MO 63301
Tel: 314-996-0109, 314-768-0109 (in IFTF)
Fax: 314-947-4261, 314-947-7689 IFTF
FOD software.

Omnicom
115 Park Street SE, Vienna, VA 22180
Tel: 703-281-1135
Fax: 703-281-1505
Publications on CCITT standards including the fax "blue book."

Open+Voice, Inc.
13771 North Central Expressway, Suite 832, Dallas, TX 75243-1017
Tel: 214-497-9022
Fax: 214-497-9028
Automated+Agent FOD/IVR system works with Novell LAN.

PM Ware
Tel: 800-846-4843, 619-738-6633
Fax: 619-738-6637
Ultrascript PC, PostScript to fax conversion utility for DOS, Windows, Mac.

Panasonic
6550 Katela Avenue, Cypress, CA 90630
Tel: 800-222-0584 voice or fax
Fax: 800-226-5329 voice or fax
Panafax and standalone fax machines.

Prometheus Products, Inc.
7225 S.W. Bonita Rd., Tigard, OR 97223
Tel: 800-477-3473, 503-624-0571
Fax: 503-624-0843
Native Wordstar, WP and Multimate file formats. ProModem, 2400MFAX,
TravelModem.

Qualitas
7101 Wisconsin Avenue, Suite 1386, Bethesda, MD 20814
Tel: 301-907-6700, Customer Service: 800/676-6386
Fax: 301-907-0905, Tech. Support: 301-718-6061
Makes 386MAX (BlueMax for PS/2 users), a Windows enhancer that allows
resident programs to operate simulatenously in multiple DOS windows. Also
makes MOVE'EM program loader that optimizes TSRs and device drivers.

Rockwell International
4311 Jamboree Road, Newport Beach, CA 92658-8902
Tel: 800-854-8099, 714-833-4600
PC modems with voice/fax capabilities. Makes chip set popular for fax modems.

SofNet, Inc.
Tel: 800-343-2948, 404-984-8088
Fax: 404-984-9956
Makes FaxWorks fax software for Windows environment.

SpectraFax Corp.
209 South Airport Road, Naples, FL 33942
Tel: 813-643-5060
Fax: 813-643-5070
Special Request, IVR with FOD; Personal Link; FaxCard; Panasonic Scanner hookup.

Speech Soft
32 Manners Road, Ringoes, NJ 08551
Tel: 609-466-1100
Fax: 609-466-0757
Low cost FOD/IVR software package that works with various voice and fax cards.

Swift Global Communications
997 Glen Cove Avenue, Glen Head, NY 11546-1599
Tel: 800-722-9119; 516-676-8000
Fax: 516-759-2000

TRT (Telephone Response Tech.)
1624 Santa Clara Drive, Roseville, CA 95661
Tel: 916-784-7777
Fax: 916-784-7781
IVR/FOD software. Demo: 916-784-7004.

Talking Technologies, Inc.
Tel: 800-945-4884, 510-522-3800
Fax: 510-522-5556
1125 Atlantic Ave, Alameda, CA 94501

Tall Tree Systems
2585 East Bayshore Road, Palo Alto, CA 94303
Tel: 415-493-1980
Fax: 415-493-7639
Fax-O-Matic, fax receiver that prints on HP LaserJet printer.

The Complete PC
1983 Concourse Drive, San Jose, CA 95131
Tel: 800-634-5558, 408-434-0145
Fax: 408-434-0145
Complete Fax, board for PC or Macintosh.

TransFax Corporation
6133 Bristol Parkway, Suite 275, Culver City, CA 90230
Tel: 310-641-0439
Fax: 310-641-4076
Fax software for LAN use.

VSI Telecom
9329 Douglas Drive, Riverside, CA 92503-5618
Tel: 800-999-8232, 714-687-2492
Fax: 800-444-8232, 714-687-2513
Markets many useful telecommunications accessories including ring-down box
which simulates two telephone lines. Useful for demos or connecting fax
machine to fax modem.

Vivitek
12493 Brookglen Drive, Saratoga, CA 95070
Tel: 408-252-6082
Fax: 408-257-9591
Fax software for Macintosh and Windows enviroments that works with multifunction fax machines. Mac product is network compatible.

Voice Interactive Processing, Inc.
P.O. Box 18387, Boulder, CO 80308
Tel: 800-826-4847, 303-442-7800
Fax: 303-938-9756
800 and 900 services including FOD, FB, IVR, and more.

VoiceFAX Information Systems
1675 West 8th Ave #106, Vancouver, BC Canada V6J 1V2
Tel: 604-732-9771
Fax: 604-732-3007
FOD and IVR services. VoiceFax demo: 604-222-8444.

VoiceLink, Inc.
1840 Oak Ave, Evanston, IL 60201
Tel: 708-866-0404, 708-467-1100 (dd)
Fax: 708-866-0412
FactsLink demo:708-866-0404 x5. IVR with FOD or verbal retrieval.

WideCom Group
55 City Center Drive, Suite 542, Mississauga, Ontario, Canada, L5B 1M3
Tel: 416-949-1866, 416-566-5259 (dd)
Fax: 416-566-0181
WideFax is the largest fax and copier, up to 24-inches wide and portable.

Xerox Corporation
201 Spear Street, San Francisco, CA 94105
Tel: 415-227-1700 for Sales, 800-822-2200 for PaperWorks Supplies
Fax: 415-227-1779
PaperWorks forms-based FOD program for Windows env.

Xerox PaperWorks Support
Tel: 800-432-9329
Fax: 800-428-3329
Press 1-sales; 2-lit; 3-tech support.

Xpedite Systems, Inc.
446 Highway 35, Eatontown, NJ 07724
Tel: 800-966-3297, 908-389-3900
FB services domestic and international. Has software for control of service by
PC.

Zoltrix, Inc.
41394 Christy Street, Fremont, CA 94538
Tel: 510-567-1188
Fax: 510-567-1280
Zofax 96/24P pocket-size fax modem.

Zoom Telephonics, Inc.
207 South Street, Boston, MA 02111
Tel: 617-423-1072
Fax: 617-423-9231
Modems and voice boards.

INDEX